P9-DVX-524

ALSO BY KATHERINE HARMON COURAGE

OCTOPUS!
THE MOST MYSTERIOUS CREATURE IN THE SEA

CULTURED

How Traditional Foods
Feed Our Microbiome

KATHERINE HARMON COURAGE

AVERY
an imprint of Penguin Random House
New York

an imprint of Penguin Random House LLC
1745 Broadway
New York, New York 10019

Most Avery books are available at special quantity discounts for bulk purchase for sales promotions, premiums, fund-raising, and educational needs. Special books or book excerpts also can be created to fit specific needs. For details, write SpecialMarkets@penguinrandomhouse.com.

ISBN / CIP TK

Ebook ISBN

Printed in the United States of America
1 3 5 7 9 10 8 6 4 2

Book design by Kristin del Rosario

PUBLISHER'S NOTE
Neither the publisher nor the author is engaged in rendering professional advice or services to the individual reader. The ideas, procedures, and suggestions contained in this book are not intended as a substitute for consulting with your physician. All matters regarding your health require medical supervision. Neither the author nor the publisher shall be liable or responsible for any loss or damage allegedly arising from any information or suggestion in this book.

To my mom and dad,
Pamela Rogers and William Harmon:

Thank you for encouraging me to ask lots of questions—
and to get my hands dirty.

CONTENTS

I contain multitudes.

—WALT WHITMAN, *SONG OF MYSELF*, 1855

The dependence of the intestinal microbes on the food makes it possible to adopt measures to modify the flora in our bodies.

—ÉLIE METCHNIKOFF,
THE PROLONGATION OF LIFE: OPTIMISTIC STUDIES, 1907

CULTURED

INTRODUCTION

We Are Not Alone

We enjoy seeing ourselves as the evolutionary apex, striding confidently, inevitably out of the primordial muck, descending from the trees, and perfecting our proud, upright posture.

I hate to be the bearer of humbling news, but we have not arrived at this perceived pinnacle alone. We've had help. I'm not talking about ancient apes or even the asteroid that wiped out the dinosaurs. I'm talking about bugs.

More precisely, we've had help from the trillions of bacteria, fungi, viruses, and archaea that have inhabited our bodies for millions of years—and had existed on this planet for billions of years before we even came along. Walt Whitman was more correct than he ever could have known. We contain multitudes beyond our wildest imaginations.

Not only do we *contain* these multitudes, but we also *depend* on them. Without our microbes, we wouldn't be here. We would never develop properly functioning immune systems, we would not get many of the extra nutrients we need from food, and our whole bodies—inside and out—would be a vast, open, welcoming landscape for opportunistic pathogens to find a home. We would be dead meat.*

Recently we have not been doing these essential microbes any favors. Through our own passion for progress and scientific crudeness, and with a dash of hubris, we've actually been doing a fairly expeditious job ruining this complex and crucial bodily ecosystem. Our microbes, known collectively as the human microbiota or microbiome,† are disappearing.

This transformation is occurring just as we are beginning to learn about our microbes' roles in health and disease. We are discovering links between changes in the microbiota and obesity, allergies, diabetes, and depression—all of which are currently soaring, despite our advances in medicine.

If we looked at our daily choices through the eyes of our microbes, it would appear as if we've been working systematically to make their lives and livelihoods difficult—if not impossible. It's as if we've been on a blind rampage against some of our most important collaborators.

There are many ways in which we've stopped providing for—and in many cases, directly assaulted—our microbes. In just the past several generations, we have dramatically changed course from the vast majority of human history and prehuman evolution.

---

* Or more accurately, we wouldn't even be here in the first place.

† We'll use the two terms pretty much interchangeably here. Some scientists have earmarked microbiome to refer only to the collection of genes comprising the microbiota—in the way the human genome is our collection of genes. But I rather like the notion, as argued by other researchers, that instead, microbiome should describe the whole environment—bugs, bile, and all—in the way we also discuss the biome of a forest or a river delta. And while we're on the topic, although it has become common to speak of the gut "flora," that is somewhat of a misnomer. Flora classically connotes plants, and in this case, we're dealing primarily with organisms in entirely different domains from plants. This could have been a fair mistake to make a few centuries ago. But even the first person to see bacteria, Antonie van Leeuwenhoek, inventor of the microscope, in the seventeenth century designated the tiny organisms as animalcules. So we'll stick with microbiome—and variations thereof.

During this time, we have been waging a war on our microbes. For example, taking indiscriminate antibiotics[‡] and, heck, just getting indoor plumbing, have upended our ancient microbiota in just the past handful of generations.[§] This assault has been a swift and efficient one—a cataclysmic extinction event in a matter of mere decades out of humans' 200,000 years. Just seconds in our species' day in the sun. And the vast health implications of this change are just beginning to dawn on us.

There is, however, another quietly powerful force acting on our microbiotas—a factor that has also veered wildly in recent generations. It is one that we are almost entirely in control of: diet.

Diet is in fact one of the most powerful ways we can influence our microbes. And we get to use it every day, multiple times. The thing is, never has what our species eats changed so quickly. Our great-grandmothers never tasted a drop of high-fructose corn syrup, let alone soda sweetened with sucralose. Several generations ago, there was no such thing as safe canning to preserve food. And for more than 99 percent of our species' time on the planet, we were all hunter-gatherers by necessity. Even the most ancient of food innovations, the invention of agriculture, has happened in a proverbial blink of an evolutionary eye.

And we are now learning that most everything we eat—from probiotic yogurt to a serving of asparagus to a fatty pork chop—has an effect on our microbes, which in turn have an effect on us. And rapidly. What you eat for one meal can change the composition of your microbiome

‡ Which only reached wide availability about two and a half generations ago.

§ Now, antibiotics, when used correctly (that is to say, not for viral infections or to make our cattle grow fatter), are a true miracle of modern medicine. They have delivered us to a life free from so many of the common killers of yore. As have many of our basic lifestyle changes that provide a radical improvement in public health. I, for one, am glad to have a working indoor toilet, to be free from hosting a four-foot-long tapeworm, and to be able to watch the snow from indoors, warmed by my home's radiators. But lately we have been taking things to the extreme.

within twenty-four hours. Not only that, but it is also becoming clear that these microbes play a key role in translating our diet to health outcomes—good and bad.

Since the time of Galileo, it has been a slow move to stretch our minds away from the gravitational pull of the human-centric universe. The microbiome provides yet another mind-boggling reminder that not only are we not masters of the cosmos, but we are also not even truly masters of our own bodies.

For most of our history, unbeknownst to us, we gave our microbes food and shelter. They gave us protection from pathogens, extra available calories and vitamins, a well-tuned immune system, and possibly even mood regulation. Changes to our genes, environment, and diet were slow. Our microbes were able to adapt to us—and we could adapt to them.

It wasn't a bad deal for either party. It is, after all, a two-way street of survival. Many species or strains of microbes have spent so long in the human gut (millennia upon millennia, getting passed along from generation to generation, back to our primate ancestors—and even beyond that) that it is the only place they can live. In other words, your microbes depend on you just as much as you depend on them. Perhaps more so. As a team of researchers noted in the journal *Nature*, "The individualized microbiota of each person has a stake in his or her fitness." If we go, they go, individually and collectively. And no one wants to be out of a home or, worse yet, wiped out forever.

So what are we to do? We can never return ourselves to that ideal, microbially perfect past.[¶] But that doesn't mean we should ignore the results of thousands of generations' dietary experimentation. After all, our human bodies and genes, for all they know, are still expecting the

¶ And there certainly was not just one single ancestral microbiota, or hasn't been since the very first human. Even that is a fraught notion, as their ancestors had microbiotas, as did *their* ancestors—back, in fact, to some of the very first multicellular organisms.

life and diet of our ancestors hundreds, if not thousands, of years ago. Without upending our contemporary lives and moving back to a lifestyle of preindustrial subsistence, we can start to pay a little more attention to that impactful role of diet.

As kimchi, kombucha, and kefir multiply on our market shelves, it behooves us then to learn more about these foods—and where they fit into a traditional diet. Many of the store-bought fermented foods we encounter today bear but passing resemblance, nutritionally or microbially, to the traditional foods our ancestors made and consumed. And as we continue to refine our meals, we should examine the major roles whole and fibrous foods play in many traditional cuisines.

Furthermore, these foods did not develop in isolation. Just as we are learning about ourselves, it is also true that no food is an island. Each is part of a whole, diverse diet—full of vitamins, protein, and fiber. Learning about these foods in their local contexts—how they are prepared, what they are eaten with, and how people incorporate them into their daily lives—provides better insights into their use across cultures. It is unlikely, for example, that simply augmenting the standard American diet** with a bottle of kombucha is going to morph your physical or psychological health into that of a fit Buddhist monk. That one might even think such a thing remotely possible in the first place is a reminder that so many have lost their traditional dietary compass (if you've ever felt adrift in an aisle of diet books, you will know yourself a fellow traveler).

This book is not here to prescribe the next fast track to weight loss or miracle health cure—in part because I do not believe such things truly exist and in part because the study of the gut microbiome is still in its infancy. Rather, it is a dip into the myriad ways we humans have

** Known, perhaps appropriately, in academic circles as SAD.

found to nourish ourselves and our microbes. It is a discovery that we must consistently incorporate these foods into our diet. It is an urging to try new things and learn to love dirty, rough foods. It is a journey in discovering a disappearing palette of distinctive flavors and vanishing traditions of handmade[††] foods.

These foods bring with them sometimes-uncanny folk wisdom as well as flavors that even the most advanced industrial processes are hard-pressed to emulate. These foods are a revelation in themselves.

To find the foods most teaming with microbial life—and most supportive of it—I traveled to their homelands, places that also happen to be known for striking longevity and hale residents. This journey involved studying and sampling cuisines from the Greek shores to the bustling streets of Seoul, from rustic barns in the Swiss Alps to the exacting cuisine of Tokyo, to learn more about how foods might help our microbes help us. And I found many delicious meals and possibly a healthier, more diverse microbiome along the way.

So, for the benefit of our microbes and ourselves, let's go see what millennia of human creativity and culture hath wrought. Let's seek out some of those traditions that are feeding long-lived people and their microbiota—keeping both alive, healthy, and maybe even a little happier.

Let's discover how to better culture ourselves.

†† In perhaps the truest sense, in that some of them depend on the introduction of bacteria from bare human hands to make their magic.

CHAPTER ONE

Microbes
In Our Guts and Under Fire

The journey to a better-cultured microbiome begins with a better understanding of just what our microbiome is—and how we have been unwittingly shaping it all our lives. Let's start this trek by taking a closer look at just how our micro-coinhabitants make their home in our bodies—and how our lifestyles have been affecting them.

Most of our microbes are concentrated in our guts.* Some of them live there full time, whereas others are just passing through. That these microbes are there at all is a recent revelation in itself. And their outsize role in our physical and psychological well-being may take yet further getting used to.

To better understand the relationship between microbes and our

* It was initially news to me that the colon is really the hot spot of our gut microbiota. Perhaps it was an eye for study marketability, funding, and general public appeal that researchers have highlighted the gut microbiome rather than colonic microbial communities.

health, however, let's start by breaking microbes into two key human-based categories: those that live permanently in the human gut and those that don't. This dichotomy is simplistic and not exactly how the microbes would see it, but it is a key distinction that is too often left out of conversations about our gut microbiota—especially as it relates to food. And it is one that leads to a lot of confusion about just what we should do with all of the new information we're gleaning about our important inhabitants.

This difference is usually glossed over or omitted entirely amid the surging enthusiasm for live fermented probiotic foods, which are those that contain strains of bacteria or fungi that have been shown to have a beneficial effect on health. Instead, we get distracted by the most inventive new kombucha flavor, the best kale kimchi, or the most local goat-milk kefir. And who can blame us? These are interesting live, effervescent cultured foods. But the microbes in our probiotic foods don't actually take up residence in our guts. They can be valuable for health, but they are not, generally speaking, replenishing an anemic microbiome. And by focusing solely on these, we are neglecting the upkeep of our full-time microbes. In fact, what our native microbes need is fiber. Complex, rustic, now-elusive fiber. These microbe-feeding prebiotics provide the food your permanent gut residents depend on to get by.

Hi, I Live Here

The microbes that reside more permanently in our guts, day in and day out, did not come from yogurt or from kimchi. They are our native microbes. These microbes are acquired at birth, throughout infancy, and in early childhood—with a few picked up here and there later in life.

These gut microbes are essential to our health and survival. They help train our immune system. They are in constant conversation with our nervous system. And they help keep the delicate balance of our guts. "These microbes evolve to their environment," says Justin Sonnenburg, a microbiologist and immunologist at Stanford University. And we may have evolved to them. He notes that "we don't just have a random collection of microbes that we pick up—we're passing microbes to each other and through generations." Through millennia of adaptive evolution, these humble intestinal microbes have come to be some of our best allies.

Who are these unseen friends? Our guts are usually dominated by bacteria from the Bacteroidetes and Firmicutes phyla,[†] which together make up about 80 percent of our microbes (though there are at least ten other phyla that make appearances). Firmicutes include the familiar *Lactobacillus* genus.[‡] Bacteroidetes include *Bacteroides* and *Prevotella*, among other genera. Another common phylum, especially early on in life, is Actinobacteria, in which the *Bifidobacterium* genus is found (which happens to be a common component of breast milk, the original probiotic).[§] These groups of microbes are not exclusive to the human gut. But some species of them can live *only* there. Our guts are their planet Earth.

Within the gut, these populations are dynamic. Most individual microbes have very short lives. So you will wake up with entirely new generations of them each morning. Some, such as members of the genus *Lactobacillus*, cycle through their whole lives in as little as twenty-five minutes. Others live and die even more rapidly. So while you were dreaming

† As a quick taxonomic refresher, the phylum level ranks just below kingdom. We humans are in the Animalia kingdom and the Chordata phylum—along with every other animal that has a backbone (kangaroos, turtles, eels). One phylum over from us in Animalia is Echinodermata (which includes starfish and sea cucumbers); far beyond that, we even see fellow Animalia phylum Porifera (sponges). So although we might tend to think most bacteria are quite similar microscopic blobs, looking at it from a taxonomic perspective, they're actually quite diverse.

‡ Which you might recognize from yogurt or probiotic supplement labels.

§ Scientists still don't know yet exactly *how* the microbes get into the breast milk; the gut is pretty far away from the mammary glands. But these extra bacteria seem to be introduced as a way to help prepare the baby's gut to digest more complex solid foods later on.

about that talking corn dog last night, your gut populations of lactobacilli could already be twenty generations beyond those you fell asleep with. That's the relative time scale between you and your ancestors who lived in the 1500s.[¶] And a lot of changes can happen in those generations of microbes. Especially if something in their environment shifts, such as an increase in pH (a drop in acidity), an introduction of a new food, a lack of the microbes' preferred fiber, or an atomic bomb of antibiotics.

Hello, We're Just Passing Through

Generally speaking, the gut isn't a naturally friendly place to microbes. Our digestive tract is a hostile environment by design. An acidic stomach helps break down food for easier digestion, but it also disarms many of the foreign organisms—from viruses to bacteria—that we come across every day. Additionally, the gut is ideally a crowded microbial metropolis, and most outsiders just can't cut it. As fermentation guru Sandor Ellix Katz puts it, the intestine "is a competitive environment. The bacteria that are there don't just move over and say, 'Oh, yeah! Come on! Welcome, neighbor!'" It's a microbe-eat-microbe world in there. All of this is a good thing for us. Only rarely does a microbe—harmful or otherwise—actually manage to endure digestion and multiply in our system.

There are, however, some microbes that can survive the harsh journey. A handful of these cause illness, such as certain strains of *Escherichia coli*. Most are probably nominally neutral. And a small fraction are actually beneficial.

Good, bad, or innocuous, though, none of these microbes are truly in our guts to stay.

¶ A time when the only ways to preserve food were to dry it, salt it, or ferment it.

I do hate to burst your highly cultured bubble. But with a spoonful—or crate-full—of yogurt, you have not actually reestablished your native gut bacteria, restoring you to peak ancestral intestinal health. No matter what the marketing will have you believe, and no matter how many live and active bacteria or strains are included. These microbes are perfectly happy biding their time in an aqueous world of lactose-filled yogurt. And they can, astoundingly, persevere through the acid-filled digestion process. But they are just not as well suited to long-term life in the human intestine.

But why not? In our attempt to supplement our diet with beneficial microbes, are we selecting the wrong ones? Even with all of our sophisticated screening techniques, are we being too narrow-minded in picking probiotic microbes? Are our busy native microbes to blame for crowding out these potentially beneficial bugs? Surely with a little more science, we can recalibrate our functional foods to contain microbes that could make the cut permanently in our guts, right?

One group of scientists undertook a clever study to find out why probiotic microbes weren't taking up residence in the gut. They started with so-called germ-free mice (those raised in a totally sterile environment with no microbes inside or out), who would have no native microbes to outcompete new microbes. Then they collected soil from a warm, acidic, microbially rich swamp, finding this location to be environmentally similar to a mouse gut, and thus a good breeding ground for microbes that would survive and thrive in those murine intestines. However, after many attempts to populate the empty guts of the mice with this rich soup of candidates, not one of the microbes survived for the long term. Even with so many potential residents and wide-open intestinal landscapes, there weren't any permanent takers. All of the microbes were adapted to live and thrive in a swamp, not in a mouse's gut, despite the seemingly similar climates.

Many of our probiotic microbes—whether from yogurt or a

capsule—are a bit like these swamp microbes. They can survive in the environment, but they aren't cut out to live there long term. They're mostly just along for the ride. Scientists have found, for example, that one to three weeks after you have consumed a probiotic, it's hard to detect any trace of the introduced species. That might be bad news if you were hoping that the yogurt you ate last month would keep your microbiome replenished for years to come.** But as scientists point out, it's not necessarily a negative that these food-based microbes don't take up permanent residence in our intestinal tracts. Perhaps we *don't* want them there forever. After all, evolution has had a long time to tailor our intestinal guest list to just the right mix. And we humans don't always get it right when it comes to messing with ecosystems.

Just because these microbes are transient, though, let's not write them off entirely. In fact, they are actually where things start to get interesting. Many of these dietary microbes get swept right through the whole system with your meals. Others might linger awhile in the small or large intestine. But while they are in our intestines, they are all still *doing* things. If you think of an individual ant passing through a picnic, you wouldn't expect it to have much of an effect. But if you imagine a full colony of ants moving through day after day after day, you might expect to see some changes. You see, microbes eat, metabolize, and excrete as they go—not just when and where they've taken up permanent residence. So any compounds a microbe consumes or produces in your intestines can change its environment—and potentially the host.†† Researchers are even beginning to suspect that a microbe's physical presence alone (thanks to the proteins on its surface) might be enough to have an impact on our immune systems.

** It is very good news, however, for those in the business of selling probiotic products.

†† That's you.

If microbes are such an important part of our health, what have we been doing to help them throughout the years? A casual survey of many of the world's traditional diets suggests that over time, groups of people—cultures[‡‡]—have adapted their diets to nourish and protect themselves by also nourishing and protecting their microbiota. When New York chef David Chang, creator of the Momofuku restaurants, speaks about creating foods with microbes, he uses words like *indigenous, native,* and even *stewardship.* These terms could come from a discussion of anything from flora to anthropological literature. Our overlooked microbial landscape and cultures deserve this kind of vocabulary as well. In fact, as we will see, they are deeply intertwined with our own places and human cultures.

After thousands of years of eating traditionally, eating culturally, science started to get in the way. In the nineteenth century, Louis Pasteur popularized the notion of germ theory, asserting that microbes (not miasmas[§§]) transmit disease. Since then, we have been diligently expunging microbes from our foods and our environments.[¶¶] And even though food-based probiotic microbes might not be as stalwart as we had once supposed, their presence in our diets can still be consequential—provided we consume them frequently.

---

‡‡ The notion of culture can refer, of course, to groups of people who share common traditions and customs (not the least of which revolve around food), but also to the intentional growing of microorganisms. In fact, for most of its history, culture, from the Latin *cultura,* meant to cultivate, in an agricultural sense. It expanded in meaning to describe human groups only in the nineteenth century. So perhaps it's time we refreshed this earlier sense of the word.

§§ Bad air.

¶¶ I am grateful we no longer have to worry about typhoid in our milk or cholera in our water supply. But our debugging fanaticism might have been taken too far—to include those foods whose microbes are helpful rather than harmful.

Our Starving Microbes

Just as we began pasteurizing with abandon, we were also ramping up other aspects of the industrial food machine. The twin blows of refining fibers out of foods and introducing a vast array of simple carbohydrates began depriving our resident microbes of their preferred food.

But without giving a hoot about these local beneficial microbes, we've pushed away thousands of years of evolved, traditional diets and placed sliced bread—or more recently, perhaps, fortified snack bars—on the ultimate pedestal. We have forgone the balance of ancestral cuisines in favor of flavor and marketing. We have cast aside millennia of slow dietary changes, instead embracing a food culture that is built to gratify our every fleeting whim. And science has been complicit, refining and concocting ingredients at record speeds to satisfy cravings. Compare our "food" options to those just a few generations ago—not to mention a thousand years ago. And perhaps you can see how some things might have gotten off-kilter in our insides.

Efficient milling, mechanized just a couple of hundred years ago, enabled the easy separation of rougher elements in grains, decreasing the fiber content of common foods, such as bread and rice. Before its advent, most human diets were exceedingly rich in fiber and other non-digestible complex carbohydrates. These are plentiful in many plants—including grains, as well as in seeds, legumes, and fruits and vegetables. These compounds feed the multitudes of beneficial bacteria living in our guts.

Most microbes inhabit a world where the main currency is the food-stuff our bodies are about to cast off. Their home is the last stop before our meals make their final exit. So, knowing what the large intestine holds, it might be a surprise to learn that the wall of this essential organ is only a single human cell thick. This thin boundary is important for nutrient and water absorption from the gut, as well as for close immune monitoring. But the integrity of the barrier (between the rest of our bodies and a tube filled with proto-feces and microbes) is obviously rather important as well. So the body has also evolved a protective mucus layer on the inside of the intestinal wall to help put some extra distance between the colon contents and the bloodstream. This layer also serves an additional purpose: providing extra food for resident microbes. It is composed of complex carbohydrates that gut microbes can feed on in the absence of dietary fiber. The body regularly replenishes this important layer to ensure protection—and microbe fodder. But when the local microbe populations don't get enough of their favorite foods from our diets, they turn voraciously to the mucus lining, oftentimes outpacing our bodies' ability to produce enough to protect the gut wall.

And this is where things get rough. Our gut wall cells know they shouldn't be coming into contact with microbes, so when the mucus is breached and the microbes make their way to the gut wall, the immune system goes on alert. If the junctions between cells are compromised and the wall loses its integrity, it becomes a more permeable barrier, allowing larger gut contents to spill out into the rest of the body. And this really doesn't look good to the immune system (as it shouldn't). Loose bacteria and food particles are not things that should be drifting through the bloodstream. So the immune system kicks in to do its job, activating cells to go on the offensive and tackle the interlopers. This creates a general climate of inflammation, which has been linked to arthritis, diabetes, heart disease, cancer, and a host of other increasingly common illnesses. Some doctors even suspect this condition—known unappealingly as leaky gut—could be in part to blame for the increase in food allergies.***

*** The theory is that as proteins from the foods, be they from peanut or wheat or egg, escape through the breached gut wall (as they rarely would have done when microbes were well fed), the immune system singles them out as threats and remains on heightened alert for them in the future.

Unclean Cuisine

Ancient granaries show that people stored foods like wild barley in large quantities at least as early as 11,000 years ago—likely even before the advent of agriculture. But not all foods keep as well as dried grains. Some foods, such as meats, can be salted. But this relies on a large supply of salt, and it doesn't always translate well to other foods.

Another way to keep food edible is to let it actually start to rot. The key is to control the decomposition—through temperature, salinity, air exposure, and in some cases, the intentional addition of certain bacteria or fungi. And so, across the globe, in much the same way that humans learned to control fire, water, plants, and animals, we learned to control what we couldn't see: microbes.

And the spoils were vast. Suddenly, with just the right curation of microbes, the morning's goat milk could last until next week. The fall's cabbage harvest could be made to last through the winter. And the fresh fish could last until next year. All thanks to microbial work that we now know as fermentation: the process of converting sugars into acids, alcohols, and sometimes, alkaline compounds.[†††] The harsh environment created by the helpful microorganisms kept harmful microbes at bay. In many cases, it even increased the nutritional potential of the foods themselves.

Just as random genetic mutations can eventually lead to the spread of well-tuned evolutionary adaptations, it's not a terrible stretch of the imagination to map out a parallel—if accelerated—action in dietary innovations. If one group of humans discovers a food source or process that allows them to stay healthier and stronger, it stands to Darwinian

††† Frequently with a by-product of gases, which speaks to the root of the word ferment as the Latin *fervere*: "to boil, seethe."

reasoning[‡‡‡] that this new element of culinary culture would spread. The benefit of food strategies is that it generally takes little more than raw ingredients and a tutorial to disseminate a new advantage. It quickly becomes a new tool in the local human quiver. First there was cooking. Then cultivation. But very soon thereafter, there arrived *culturing*—of the fermented variety.[§§§]

Through millennia of experimentation, the foods that didn't make people sick stuck around. Fine-tuned over seasons and generations, they provided a longer-lasting food supply and more consistent, reliable access to calories and essential nutrients through the seasons.

Beyond basic nutrition, these fermented foods also delivered regular doses of a wide variety of microbes. Many scientists now think that our bodies actually came to expect frequent servings of microbes—which performed similar functions to our own native microbes, augmenting their work and bringing other passing benefits.

These days, though, we are distracted by our conflicting impulses to satiate our desires (for the sweet, the salty, the caloric) and to follow the latest health-promising headlines (choosing the sugar-free, the low-fat, the low-carb). But in all of these choices, we've unwittingly neglected those trillions of organisms that have evolved alongside (and *inside*) us and that have helped keep us healthy.

And not only have we been making the wrong food choices, but we have also been making the wrong lifestyle moves.

---

‡‡‡ Such an explanation has indeed been made for the spread of dairy husbandry and some groups' ability to continue comfortably digesting dairy into adulthood.

§§§ We'll use both culture and ferment to loosely describe the process of microbially transforming food. Culture can also refer to a specific group of microorganisms introduced to grow in a medium, as in a starter culture—for cheese or sourdough bread, for example.

Sanitized

Just as our diets would hardly be recognizable to our ancestors, our lifestyles have transmuted into something utterly unnatural—and far too clean. Humans and animals are not the only things with microbiomes. The environment has them, too. Soil has its resident microbes, as does seawater, as does your home and even your coffeemaker. For most of human evolution, the external microbiotas we came into contact with were those belonging to the familiar, natural world whence we also came. Even for most of the time we have been building shelters, the floors have been dirt. There were no sealed windows, no HEPA filter vacuums, and certainly no antibacterial cleaning products. We lived a life surrounded by the vast microbiota of nature. All that has changed in the past couple hundred years, and with dramatic results.

We have long known that kids who grow up on farms—in regular contact with soil, animals, and all matter of material city folk might consider dirty—are less prone to overactive immune systems that can spur allergies and asthma. But particularly since the acceptance of germ theory a century and a half ago, we have been cleaning more and more each year. The initial benefits were myriad. Residents of London, for example, stopped getting cholera from their local public water supply. Surgical patients can now be confident that they are unlikely to contract gangrene from an unsanitized scalpel. And most of us will not endure a disabling guinea worm infection. But we took a positive thing too far, with chemistry and microbiology—and marketing—pushing us past the point of diminishing returns. Now we might be *too* clean, our bodies expecting those many environmental microbes that once kept our immune systems balanced. This notion has gained such traction in the scientific world that it has come to be known as the "old friends" hypothesis—the idea that we are now missing those important organisms we'd devoted so much energy to eradicating.

There is also a limit to our cleaning power. Even when we try to strip all microbes from our homes and buildings, environmental microbes can come back with a vengeance. Studies have shown, for example, that hospitals with closed air systems have higher rates of circulating pathogenic bacteria than hospital wards that open their windows to the outdoors.¶¶¶

Unlike the wilds encountered by our ancestors, most of the environments we inhabit today don't have much to give. Often, *we* are the largest source of microbes in our spaces. Scientists have found, for example, that when we arrive in a clean hotel room, it takes just a matter of hours for us to colonize it with our own personal blend of bacteria. As Maria Gloria Dominguez-Bello, a Rutgers University microbiota researcher, says, in the jungle, the floor colonizes us, but in the built environment, we colonize the floor. The microbial equation has been flipped. To make matters worse, we have also become impoverished couriers.

Historically, a baby was born to a world without sanitizing soaps and antibacterial rinses. It passed through the birth canal and into the arms of another (unscrubbed) human. It was not washed with soap or whisked away into a nursery ward to sleep in a plastic bassinet. As products of the Pasteurian age, we now begin life in microbial poverty. To be sure, a certain amount of cleanliness can save babies—and mothers—from infection. But as with many other aspects of twenty-first-century life, these practices might be interfering with a long-established microbial balance.

The first moments of life play an important role in shaping the early trajectory of the microbiome. When an infant is born naturally via the birth canal, the first microbes it meets are residents of the mother's vaginal microbiota, which shift over the course of pregnancy in anticipation of this very moment. These first microbe colonizers are often mixed with some from the mother's gut via normal fecal contact. This mode of arrival populates the infant's skin and mouth (and thus the

---

¶¶¶ A feature Florence Nightingale might have been onto in the 1850s, when she recommended patients do best in rooms with open windows.

digestive tract) with the same brew of microbes that has been the first bacterial bath since time immemorial. This special cocktail of microbes provides a rich, diverse, and carefully tuned complement of functions that support the baby's digestive, immune, and other systems in its first weeks, months, and years of life.

Babies delivered via cesarean section, however, first encounter skin microbes from medical staff and new parents, as well as microbes from the ambient hospital environment—not vaginal or gut microbes. These children have an entirely different collection of founding microscopic partners. We have taken this new route into the world more and more. Just forty-five years ago only about 5 percent of births in the United States were cesarean. In less than twenty years, that figure jumped to nearly one in four deliveries. In some Latin American countries, that rate has climbed to nearly half of all deliveries. To be sure, some of these procedures are medically necessary, but many are not.

The populations inherited at birth live on with us for months, perhaps years, to come. Dominguez-Bello says she can tell how a baby was delivered at the age of one month simply by analyzing microbes from a diaper sample. Even one *year* out, she says, a swab of the baby's skin provides an indication of birth method.

Even for babies delivered vaginally, their microbial inheritance is not what it once was. Research has found that mothers, themselves now microbially impoverished, are also passing along less robust microbiota to their children.

Firebombing

What microbes we do still inherit are also under serious threat. Americans are now prescribed some 258 million courses of antibiotics each year. With a population of 318 million people, that equates to roughly eighty prescriptions for every hundred people. Every year. Many of these prescriptions are for broad-spectrum antibiotics, which work indiscriminately on many types of bacteria. This method of attack is helpful in cases when a doctor might not know (or care to find out) the exact type of offending bacterium and can instead wipe them all out with a pharmaceutical flamethrower. But the casualties of these drugs do not stop at harmful microbes. They also include innocent bystanders, many of which are highly evolved to our guts—and very difficult to replace once lost.

Certainly, many antibiotic prescriptions are necessary. But many are not. In Sweden, the antibiotic prescription rate is closer to thirty-eight courses for every hundred people annually, and the population seems to suffer no ill effects. (In fact, some might argue that they are better off for it.) Any ailment caused by a viral infection—among the most common reasons to see a doctor during cold and flu season—will not be helped one iota by an antibiotic. Antibiotics work by destroying bacterial cells, stopping their repair mechanisms or other means specific to how bacteria are built. Using antibiotics in hopes of vanquishing a viral infection is like expecting kryptonite to work against the flying monkeys in *The Wizard of Oz*.[****] In the case of a virus, a round of antibiotics could in fact actually make the body more vulnerable to infection. Yet patients, impatient to feel better, often request treatment of some kind over waiting it out. And treatment usually means antibiotics. (There are antiviral medications for some of these infections, but the best thing for a run-of-the-mill virus is usually hydration and rest—often not a popular option in our fast-paced, treatment-driven culture.)

**** Indeed, these drugs should more appropriately have been labeled *antibacterials*. Although the *antibiotic*—or "against life"—moniker might turn out to be more appropriate and more broadly applicable than intended.

Interestingly, within the United States, prescription rates are not uniform. They vary widely by state, with far higher rates in the South (with nearly one antibiotic prescription per person per year) compared with the West (where closer to six of these prescriptions are written for every ten people each year). This pattern also curiously mirrors the geographical rates of obesity. Of course, correlation is far from causation, but farmers have known for decades that feeding continuous low-dose antibiotics to livestock makes them gain more weight faster. We are only now starting to wonder about a similar connection for us.

We know that these drugs can wreak havoc on the adult human gut microbiota. But what about those of kids, which are just getting established—and which might also be intimately involved in shaping their developing immune and nervous systems? This is an important question. Because children are receiving a lot of antibiotics.

Children in the United States get on average twenty courses of antibiotics before their eighteenth birthday. And the largest percent of these go to children before their second birthday: some 1,300 Rxs for every 1,000 babies. Not to mention the dose many babies and their mothers get at birth as a matter of course. Some kids' microbes bounce back, but one study found that other children's gut bacteria were still out of whack *four years* after finishing a single course of antibiotics. Nevertheless, some 70 percent of children who visit a doctor for an upper respiratory infection leave the appointment with an antibiotic prescription. This, despite the fact that more than 80 percent of these infections are caused by viruses, not bacteria. This math doesn't look good for the gut microbiome—or for other related aspects of health.

Researchers have found that children who were prescribed antibiotics before six months of age or multiple times during infancy were heavier based on their height by age two. And antibiotic-altered gut microbes might impact allergy risk as well as behavior in kids. Scientists are finding that antibiotic use in mothers can impact children before birth—and even before pregnancy—by altering the microbes the mother possesses and thus the microbes that child is able to inherit.

Many other aspects of our daily lives that we hardly give a second thought can also change our microbes. Anything that alters the environment of our guts can force out some species while encouraging others. Proton pump inhibitors and other antacids, for instance, might provide temporary relief from heartburn and acid reflux, but by lowering the digestive tract's acidity, they also lower its defense against invaders and its habitability to locals. One of the reasons many gut bacteria are beneficial, in fact, is *because* they produce acids. By doing this, they lower the pH of their surroundings, making it more difficult for many would-be pathogens to survive. And research shows that acid-neutralizing medications are indeed linked to reduced gut microbial diversity and higher risk for infections, such as pneumonia, as well as to vitamin and mineral deficiencies (acid helps break down food, separating out these crucial compounds for easier absorption).

Additional lifestyle factors including the use of infant formula, increasing chronic stress,[††††] smoking,[‡‡‡‡] and lack of exercise are also conspiring to change the microbiome.

Diversify, Diversify, Diversify

A key change we're seeing in the microbiome related to health issues is a lack of diversity.

In any ecosystem, diversity is an indicator of health and resilience.

†††† Or short-lived stress. One study found that compared with earlier in the semester, during exam week, college students experienced a dip in lactic acid bacteria present in their guts.

‡‡‡‡ Among many other factors, smoking is also linked to lower gene richness in the microbiome—and quitting smoking leads to a recovery of diversity.

A diversely populated meadow or forest will more easily fend off invasive weeds and cope with change than a simple monoculture.[§§§§] Numerous studies have found that people eating traditional diets and living traditional lifestyles harbor a far more diverse lineup of microbes in the gut. For example, groups living largely apart from globalized society—from the jungles of the Amazon region to the highlands of Tanzania—have at least a 30 percent higher diversity of microbes living in their gut than folks living in upper-income regions. One Amerindian group known as the Yanomani tribe had gut microbiotas that were closer to 60 percent more diverse than the average American or European gut. In fact, if we map the diversity of the gut microbes of these residents of South American and that of African indigenous people on a scatter plot against those of "healthy" North Americans, the two indigenous groups cluster more closely together, despite the vast geographic distance between them. In this context, those of us living in so-called developed societies appear to be the impoverished outliers.

Even within less-enriched guts, though, there is still a telling range of diversity. Looking at a broad population of individuals from wealthy countries, researchers were surprised to find that diversity levels did not follow a typical bell curve trend, which would show most people having an average number of microbe species, a few having a very low number, and a few having an exceptionally high number. Instead, it looked more like a saddle, with two large camps of people on either end: those with relatively low diversity and those with relatively high diversity (with a stunning difference of about 40 percent between the two main groups). Those in the lower-diversity group were more likely to be overweight or obese. Separate studies have also shown that people with inflammatory

---

§§§§ Hence the heavy use of intensive chemicals for maintaining most of our large-scale monoculture crops; nature never intended sprawling acres of identical corn plants.

bowel disease (such as ulcerative colitis or Crohn's disease) have some 25 percent lower diversity in their gut microbiotas than healthy individuals. Low microbe diversity has also been linked to inflammation. Chronic inflammation can create an environment that selects for a community of microbes that in turn further increases inflammation, creating a vicious cycle that may increase the risk for many diseases, including arthritis, dementia, and certain types of cancer.

Among those who are overweight, microbial diversity might also be an underlying indicator of health. Looking at a small group of obese patients in France, researchers found that in similarly overweight individuals, those who had the most healthful diets (marked in their criteria by low consumption of sugar and a high consumption of fruit, yogurt, and soups) had a more diverse microbiota than obese participants of similar weight who had the least healthful diet (with lots of sugar and few fruits or yogurts). In studying these two types of obese profiles (low- and high-diversity), researchers also found that people with low microbial gene counts (a mark of low diversity) were more likely to have pro-inflammatory microbes, whereas those with high microbial gene counts had different, more anti-inflammatory species, such as *Faecalibacterium prausnitzii* (which might be protective against some diseases of the gut and beyond). The team also discovered that a low microbe gene count in these patients predicted insulin resistance (a precursor to diabetes) even better than weight did. A further study found that helping either type of obese participant (low-diversity or high-diversity) lose weight with an improved diet boosted the microbial diversity. "These findings support the reported link between long-term dietary habits and the structure of the gut microbiota," writes one team of researchers. "It also suggests permanent adjustment of the microbiota may be achieved through diet." These studies go to show how crucial a microbe-friendly diet is for long-term health.

Overnight Changes

The good news in all of this is that we do have a fair amount of say about our microbial communities. We can choose a lifestyle and diet that will shift the microbiome in a better, healthier, more diverse direction. And these changes don't take weeks or months to have an impact. Thanks to the short life spans of bacteria, a change in their environment can mean large changes in populations and functions, literally overnight.

This stunningly quick shift was documented in a well-controlled human study performed by Harvard University researchers. Peter Turnbaugh, a microbiologist now at the University of California, San Francisco, and his colleagues recruited ten people to participate in an investigation of diet's impact on the microbiome. They had previously found distinct changes in mouse microbiota when the rodents' diet was altered.[¶¶¶¶] So it was time to find people willing to eat the same thing for five days straight.

Turnbaugh and his colleagues divided the group in half: one half ate a diet entirely composed of animal products (meat, eggs, cheese, etc.) and the other half a high-fiber plant-only diet (legumes, grains, vegetables, fruits, etc.). Then the groups went through a waiting period before trading dietary regimes. The participants, who came from a range of dietary habits—including one lifelong vegan—started with wildly different microbiota. During the study, the researchers could still identify each individual based on their microbiome, but their microbe profiles and those microbes' active genes quickly began to conform based on

---

¶¶¶¶ Mice are still the most common animal model for microbiome studies—not least because you can now mail-order germ-free mice. But they are also problematic. You see, mice (and, actually many other animals) have a charming tendency known as coprophagy, the tendency to eat feces. Consuming your or your cage mate's feces might provide additional nutrients. But it also provides a wealth of gut microbes, raising the possibility that outcomes seen in mouse studies are far stronger than they would be in humans, thanks to these, ahem, *reinforced* microbial changes. And our aversion to such supplementation.

their assigned diet groups. In the end, they found that our microbiome is astonishingly adaptable to our diet—for better or for worse. All the more reason to keep our microbes in mind when we consider what to have for our next meal.

Before we dig our forks into the wild world of microbe-feeding food, though, let's get to know a little bit more about these essential partners who call us home and shape our health.

CHAPTER TWO

What's in the Gut

Most of us go through our lives thinking of ourselves as an individual, singular human. Whether we realize it or not, however, we each have hundreds of species that call us home—at least 100 to 200 in the gut alone. And we are each, at any given time, hosting some 40 trillion individual microbial friends—roughly 6,000 times the number of people on the planet. The body turns out to be a pretty dynamic and populous place.

In general, these microbes are smaller than our human cells. Altogether, they round out to just a few pounds, mostly tucked away inside the large intestine. Although they are small, they are busy. Each microbe, in most cases, is occupied in creating cell walls, making RNA, producing enzymes, and carting around little energy factories, like a microscopic Richard Scarry cartoon. Despite their small individual size, collectively, they comprise a marvel of activity and potential.

When we talk generally about the microbiome, it is often shorthand for the microbes that live in the digestive tract—primarily the colon. But separate human microbiomes also exist for the skin, mouth, nose, and other areas of the body. Each of these regions has a distinct and fascinating ecology. And each biome has a radically different environment. Just as we would not expect to find the same plants and animals living in the ultra-arid Atacama Desert and in the humid Amazonian rain forest, we have also discovered that vastly different populations of microbes live on our shoulders than in our armpits. We even have distinct communities on our left and right hands. Our bodies are a crazy, diverse jungle—and desert.

A Brief Tour

Before we can understand the microbe's role in our guts and our health, it helps to get oriented with the more macro story of our digestive tract. So let's begin with a meal.

The story of food, once it leaves your fork, may not be a pretty read, but it is every bit as astounding—and as full of twists and transformations—as a compelling novel. From beginning to end, a bite of dinner travels some thirty feet over the course of a few dozen hours or more. All along the food's way, the intestinal tract is communicating back to the rest of the body via nerves, neurotransmitters, hormones, and the immune system. Muscles are coordinating in a masterful progression of squeezes, sloshes, and releases to move the plot along.

Although we think of our digestive tract as very internal (and very personal), it is in fact a kind of opening to the outside world. It is a tube through which foreign things (most of which we hope are edible) pass every day. A sort of second skin, it is highly absorptive, allowing us to extract the nutrients we need. To maintain order and health, the immune system is ever vigilant, searching for and battling potentially harmful elements that might have made it down the hatch during the fifty-plus hours it takes your dinner to complete the trip.

It all begins with a bite. Let's make this a forkful of potato salad. After succumbing to the mouth's mastication* and a little action from enzymes, which break apart chemical bonds to cut food into yet smaller bits, our bite is swallowed and pushed down the esophagus. At the bottom, it meets a sealed door. This sphincter opens to allow the food, now known in technical circles as a bolus, into the acidic stomach before closing again to keep that searing stomach juice from burning the esophagus's delicate mucus. (When stomach acid does get into the esophagus—for reasons that might sometimes have to do with your microbes—the result is unpleasant acid reflux.)

The stomach is a harsh environment, with a pH of around 2.5 (equivalent to a pouch of hot, churning white vinegar). These gastric juices†—along with the mechanical contractions—break down most of the remaining food chunks here by chemically incinerating the bonds that once tied them together. After about three to five hours, what was once that picnic side dish (but is now chyme) exits the chaotic stomach and enters the lazy river of the intestines. In the small intestine, more digestive juices further deconstruct particles of food, and most nutrients are absorbed into the body. After getting squeezed along this convoluted conduit for twenty-some feet of digestion and absorption, what is left of the potato salad travels into the large intestine as mush. This slurry spends its last hours inside the body here in the large intestine, where moisture is extracted prior to expulsion. But that is not all that's happening in there, as we once thought.

Enter the microbes. The digestive tract is populated with microbes

* What food we're eating actually determines the size of the particles we chew it into. One study found that people chewed carrots to about the size of 1.9 millimeters, Emmental cheese to 2.4 millimeters, olives to 2.7 millimeters, and German-style pickles (gherkins) to about 3 millimeters.

† The juice, of which the stomach can produce some two to three liters each day, is a special blend of hydrochloric acid, mucus, and enzymes.

from top to bottom. They are relatively few and far between in the stomach (once thought to be too acidic for any life, it has now been shown to harbor some life, such as the sometimes damaging and sometimes beneficial *Helicobacter pylori*). A bit more appear in the small intestine,‡ but as we have learned, the majority of them make their home in the large intestine.

Our teeming large intestine is the last chance for our bodies to extract—by way of our microbes—any last useful elements from our meals. For most of our species' time on Earth, this was incredibly important, given there was far less food security. Many colon-loving microbes give us extra essential vitamins, such as vitamin K, and improve mineral absorption, such as that of calcium. And depending on your microbial makeup, they can extract as much as 15 percent more calories from your food. Mice raised in the lab without microbes eat more food and gain less weight than normally colonized mice. Once given gut microbes, the mice easily achieved a normal weight—while eating less. A great advantage if you're foraging for food in the wild. Now we might be inclined to tell those little buggers to get lost (think of all that extra potato salad you could eat!). But as we will see, getting rid of these microbes wouldn't end well.

In order to learn more about what microbes are providing these services, scientists started at the end: sampling the material that makes its natural exit.§ However, "while it is relatively easy to obtain fresh fecal samples, the information obtained from them does not represent the

‡ More scientists are turning their attention to this hard-to-study region of the digestive tract. Studies in animals have shown this area to house a very different population of microbes performing very different functions than those in the large intestine. Overgrowths of the wrong bacteria here can cause serious health issues. And for many probiotics that seem to be all but missing from the highly competitive large intestine, this might be an area where they have the opportunity to be more effective.

§ The more appetizing stuff comes later, I promise.

complete picture within the gut," notes one team of researchers writing in the journal *Gut*. As we have seen, the small and large intestines play vastly different roles in our bodies—the former primarily absorbing nutrients, the latter water. And "we know that the small intestine contains a very different abundance and composition of bacteria," they write. Other researchers frame their research tactics another way.

"Most of what we study is poop," declares Jonathan Eisen, a University of California, Davis evolutionary biologist studying microbes, when I meet up with him for a chat in a local park on a sunny spring day. He sports a beard and one of his signature science-themed T-shirts (today's is not, alas, his famous *Ask Me About Fecal Transplants* one). "Are fecal samples good?" he asks. "Yeah, they're great. We've learned an enormous amount from them. But they're a mix of everything that's in the system," he says. He likens trying to exhaustively map the whole intestinal microbiota via fecal samples to trying to understand all of Australia by studying the outflow from one river. Sure, you can collect and examine the feathers, shells, bones, and chemical compounds that come out, but you're not going to have a very good sense of what's really going on in there—the crucial role cassowary birds play in spreading banana tree seeds, or whether or not dingoes and monitor lizards compete for the same habitat. So rather than relying on the "river" output, he says "it would be better to actually go *into* the system and catalogue things in their native environment." But doing that in humans is tricky. "Very few people have done the internal sampling. We really don't have a good idea for the whole biogeography."

Scientists have nevertheless been able to find some trends in these outflows, he says. "If you feed a hundred people the same food compounds, on average, similar things happen to the microbes in their poop," Eisen says. But the emphasis is on the word *average*. And we know that everyone is different, as are their microbes.

Meet Your Microbes

Just who are these fellow travelers? We might be inclined to imagine that the lowly microbes who are content to set up shop in our colons are a kind of bland, amorphous band of brothers, but no, as we shall see, they hail from wildly different evolutionary lineages. When you think of microbes, perhaps one of those colorful microscopy images pops to mind. Or just a nebulous notion of some unseen creepy-crawlies. The truth is, when it comes to understanding what these organisms look like and what they do, saying *microbe* is about as descriptive as saying *animal.* In fact, it's even less useful than that.

You see, we carry around all three *domains* of life in our guts: bacteria, eukaryotes (multicelled organisms), and archaea (other single-celled organisms).[¶] Viruses also call our guts home (but are debatably alive since they can reproduce only by commandeering another organism's cells). Each of these domains in turn includes entirely different kingdoms and phyla. So within the bacterial branch[**] alone, there may be organisms essentially as (if not far more) genetically different from one another as we are from moss, or a Pomeranian is from an amoeba. All of *us*—humans, Pomeranians, amoebas, and mosses—live under the eukaryote domain. Most of the eukaryotes in our guts are fungi. Historically, we were also home to numerous more macro, multicellular companions, such as parasitic worms.

Not only is your own microbe collection incredibly diverse, but our microbiomes overall are also strikingly different from person to person. Whereas you share 99.9 percent of your human genes with your neighbor—as well as someone living halfway around the globe—your

---

¶ Distinguished from bacteria as a separate domain of life only in the twentieth century.

** Bacteria were the only (fully fledged) life-forms on the planet for a few billion years.

respective microbial communities might overlap only a tiny bit.[††] And researchers are now finding that even a single species of microbe in your gut could share less than half of its active genes with a microbe of the same species in that neighbor's gut.

These genes are a big component of a microbe's activity, and by extension, its influence on the environment and on us. Activated genes determine how busy a microbe is, what it is inclined to eat, and how it communicates—or does battle—with other microbes. As a whole, the microbiome is a genetic powerhouse. Each of our microbiomes holds about a hundred times more genes than our own individual human genomes. That means we each carry around many more microbial genes than human ones. So if you were still inclined to think of yourself as just a human individual, it might be time to rethink that old-fashioned notion.

But who exactly are these residents of ours, these faceless microscopic masses? Over the years, scientists have developed various ways to classify these organisms. For example, bacteria come in three different shape-based categories: bacilli, cocci, and spirilla. Bacilli are the familiar rod-shaped bacteria. Microscopic images of these bacteria often look like textured pill capsules or long fuzzy sausages. Bacilli include everything from *Lactobacillus plantarum*, present in fermenting sauerkraut, to *Bacillus anthracis*, which is behind anthrax. So you can't always judge a bacterium by its cover. Cocci are more sphere-shaped and tend to be especially tiny. They range from *Streptococcus thermophilus*, an

†† Although, as in a lot of health research, these statistics might be most descriptive of those living in wealthy nations. Research into the gut microbiota of indigenous populations of Papua New Guinea, who have been living in the same location, eating the same foods, and living communally for generations, suggest that they actually have relatively similar microbiotas. "Whereas the average person in the United States, we live in this sanitized world, so we've all enriched our own type of microbiome," says David Mills, a molecular biologist at the University of California, Davis. So the broad assertions that each person harbors a wildly different microbiota may be more "a result of the last three hundred years—or even one hundred years . . . and [perhaps] we used to actually all have much more similar microbiomes," he notes.

essential ingredient in yogurt, to *Staphylococcus aureus*, which causes staph infections—including the dreaded MRSA (a form of drug-resistant staph). Finally there are spirilla, which, if you hadn't already deduced, have a spiral (or rather, corkscrew) shape. These are a bit less important, so far as we know, in the disease and food worlds, but it is nice to know that there are some wild and crazy bugs out there.

The microscopic fungi in our guts and our food have an even wilder array of forms than the bacteria. Some, such as *Saccharomyces*, which are important in everything from baking to brewing, may have an oval or egg shape. *Geotrichum*, a genus that is present in some yogurts, looks much like the rod-shaped bacilli bacterium. And then there is *Aspergillus*. This brand of fungus, cultivated to ferment soybeans, sprouts an ostentatious head to disperse its spores. In its full glory, it looks something like a flashy ornamental garden allium.

All of these shapes and features help to determine how microbes assemble, reproduce, and move. Most bacteria reproduce by mitosis, essentially splitting in two: generating duplicates of the essentials, swelling in size, and then cleaving down the middle, creating two (theoretically) identical cells. Fungi reproduce in a variety of ways, including in the above manner and via spores. Fungi are stationary creatures, whereas many bacteria use a thin tail, known as a flagellum, to propel themselves around and hairlike structures known as pili to adhere to surfaces.

• • • • •

The first step in understanding how all of these microbes impact our health is to find out exactly who is there—in our guts and in our food.

The cast list of microbes is still incomplete and is likely to stay that way for quite some time. While scientists have been busy the past few centuries cataloguing the stunning diversity of plants and animals in the world, our microbes have gone largely unstudied. For one, they are invisible to the naked eye. To boot, most of our gut microbes are anaerobic and don't take well to the oxygen-filled world we (and our laboratories) inhabit. For much of scientific history, we had to rely on organisms grown in petri dishes for our research into microbes. Not surprisingly, taking a swab from one environment and seeing what will grow in a totally different environment is not the best way to measure who actually lives in that first environment. Think about trying that with plants and animals—taking a sample from the Amazon River to see what would flourish in the Atacama Desert. What did survive and thrive in the dry sand would hardly be a good representation of the original species' abundance underwater. But for decades, that's basically the best we could do, and consequently, we had no idea just how rich and diverse our microbial rivers really were.

Now we have genetic sequencing, which can tell us who was living in an environment even if they don't survive in the lab. With the increasing speed and dropping cost of genetic scans, we can give samples a quick run through the sequencer and get a real tally of who is in the mix—alive or dead. Researchers can look for differences in a specific signature known as the 16S rRNA gene. Our version is different enough from the type bacteria have that it's a shorthand way to sort bacterial from human genetic material. Plus, it's different in each species, so scientists can now match 16S rRNA sequences to known lists of species without having to assemble full, exhaustive genomes for each individual. Zip-zap, microbes tallied. It might not be the perfect, ultimate solution to understanding our microbe populations and all of their dynamics, but it is far more comprehensive than attempting to grow microbes from a sample in the lab.

The information from genetic studies has led to all kinds of amazing insights about the gut. But even when we think we see a promising link between microbial patterns and health, it is important to take a second and a third look.

Case in point: North American and European microbiomes are usually dominated by bacteria from two phyla: Bacteroidetes and Firmicutes, with Bacteroidetes as predominant. For several years, a number of small studies—in lab animals and in humans—were turning up an interesting pattern: Bacteroidetes and Firmicutes ratios seemed predictably skewed in obese individuals. The lean tended to have more Bacteroidetes, but those with extra weight had a disproportionate abundance of Firmicutes.

Ta-da! Solving dieters'—and the world's—weight problem might be as simple as slanting back this ratio into slimmer, Bacteroidetes-filled territory. Diet gurus pounced on the formula.

Alas, such a simple prescription is probably not all that useful. In fact, a broader survey of data showed there was actually no statistically significant link between a person's BMI and his or her ratio of Bacteroidetes to Firmicutes. Besides, could as broad a taxonomic category as a *phylum* really determine our weight? Would you say, "Oh, I left my newborn at home—but it's okay, there are a couple of vertebrates watching her"? It might matter precisely what type of vertebrates you had babysitting your offspring: grizzlies, guppies, or human grandparents. Or if your doctor recommended switching to a plant-based diet, you probably wouldn't rush out to find the nearest hemlock root to gnaw on (unless you really, really couldn't imagine life without bacon). So we shouldn't expect members of bacterial phyla to be so much more uniform than those we encounter in the macro world.

Cautionary tales like this have led us to more nuanced distinctions of our populations of microbes. As we've seen, beyond the phylum level (and class and order and family), there are many, many genera‡‡ in the gut—some exotic and poorly studied, but others, such as *Lactobacillus* and *Streptococcus,* familiar and well-examined subjects. Even these two examples, though, are a reminder of why it is crucial to drill still further down, at least to the species level.

Both of these genera are members of the Firmicutes phylum. They are both lactic acid–forming bacteria. This creation of lactic acid, in food-based species and in gut-based species, is a beneficial service (fermenting food and keeping the gut acidic, respectively). But whereas *Streptococcus thermophilus* helps make yogurt—and a spread of other delicious fermented foods—*Streptococcus pneumoniae* (which causes pneumonia) and *Streptococcus pyogenes* (which can cause everything from strep throat to flesh-eating diseases) can lead to illness and even death. As one Greek researcher puts it, we ought to think of the spectrum of microbe behaviors as we do other forms of life. A Pomeranian is a highly (one might say excessively) domesticated companion animal, but a wild gray wolf can (and may well) kill you—even though both are members of the genus *Canis.* The same can be said for members of the *Streptococcus* genus—and many others.

Still, to *really* get to the bottom of how these microbes—both residents and passers-through—work, we must drill down further even than species, to the individual strain. For example, plenty of *Escherichia coli* strains are harmless, residing in our guts without ill effect, and some can even be beneficial, such as the probiotic strain *Escherichia coli Nissle* 1917.§§ However, other strains of this species, such as *Escherichia*

‡‡ Genera, too, can be a broad category. Consider the genus Homo, which includes us, the Neanderthals, and others all the way back to early hominins, who were just beginning to walk upright and might not have had language.

§§ Isolated in 1917 from a World War I soldier who managed to stay healthy during an epidemic of diarrhea.

coli O157, can kill us. So specifying a particular strain in an experiment or a food is crucial as we try to better understand the impact microbes can have.

All of this is achingly complex, and the data incomprehensibly vast. But even just scratching the surface of this rich and complex world is yielding big payoffs for health.

Big Roles for Small Bugs

Some people are keen to hail the microbiota as the "forgotten organ." But I think that isn't giving the microbes enough credit. As we've seen, they do help us process some of our food and extract more calories and minerals from it. But as we will soon see, they do so much more. They also talk to and tone our immune system. They help regulate hormones. And they are part of our intricate neural communication system, which governs our moods. All in all, they're hardly the three pounds you'd want to shed. (Although treating them well might be able to help you shed at least that much in actual ballast.)

Microbes might be linked to weight gain and obesity in more ways than simple energy extraction. Although altering the ratio of Bacteroidetes to Firmicutes might not be the silver bullet it was once hailed to be, there are other trends lurking in our guts that correlate with being overweight (a condition that now affects more than 160 million people in the United States and some 2.1 billion people worldwide). Researchers have found that giving germ-free mice the microbes of an obese person tends to make those mice obese as well. One reason for that is that microbes might affect the hormones that make us feel full or hungry. Other research suggests they could impact how we metabolize foods and even medicines. This finding might account for some of the unexplained variation in drug responses from person to person. Still other microbes might combat harmful conditions that arise with obesity, such as inflammation. And microbes that help strengthen the gut lining could reduce the amount of harmful compounds that leak out and cause low-grade inflammation in fat tissue.

A similar balance of microbes can also be tied to wellness or illness in different people depending on their background and diet. For instance, *Prevotella* bacteria are associated with plant-heavy non-Western diets in healthy individuals, including many Africans who eat a traditional plant-centric diet. But *Prevotella* are also often prevalent in the guts of Western patients with AIDS, where it is seen as a symbol of dysbiosis—a microbial imbalance.[¶¶]

Our Microbes, Our Immunity

The immune system is charged with protecting us from invaders. Microbes are organisms that are definitely not genetically *us*. Yet the immune system tolerates (and even depends on) a vast number and variety of these organisms for proper development. So the dance between the microbes and the immune system that has evolved is a fascinating and delicate one.

It is now commonly accepted that part of the reason for recent increases in allergies—as well as possibly some autoimmune disorders—is that we've made our external environments too clean. There's no reason for our bodies to fight pollen, but for many people, the immune system

---

¶¶ Justin Sonnenburg and his research partner and wife, Erica Sonnenburg, at Stanford University are quick to point out, however, that "because our definition of a healthy microbiota comes from studying Americans and Europeans, it's likely that our view of what is normal is highly distorted." Additionally, as a team of Japanese researchers notes, "It is unclear whether normabiosis is similar in healthy individuals and between Western and Asian people." Clearly even these basic terms are still in need of a lot more research.

spots those tiny particles and goes on the offensive, spurring inflammation, mucus, and watery eyes. The hygiene hypothesis originated from studies in the 1980s of children who grew up in large families. These kids, especially the youngest siblings, were far less likely to develop allergies than their peers who grew up nearby in single-child households. Scientists hypothesized that the exposure to many more microbes and particles in the environment (specifically from germ-covered siblings) at a young age primed the children's immune systems to attack truly harmful invaders while taking a measured approach to less threatening compounds, such as pollen or peanut proteins. Or, as I have thought of it, perhaps if my dog had met a real burglar or two as a pup, he wouldn't bark so ferociously at the friendly mail carrier every day.

So generations of kids (including me) have been sent outside to play in the dirt so that we might encounter more foreign particles, more exotic microbes to expand the vocabulary of our immune systems at an early age. Interestingly, more recent research suggests that our *internal* microbes might play an even more important role in developing a healthy immune system than these external ones. And that the exposure—or lack thereof—to these internal microbes is happening at a crucial time. Recent studies show that there seems to be an important developmental window during which a perturbed microbiota (disturbed by substantial early antibiotic use, for example) could have lifelong effects, increasing risks of immune-related ailments, such as asthma and allergies.

A jumpy immune system can do more than rev up in response to harmless foreign particles. It can also turn on the body itself. Researchers are currently investigating connections between the gut microbiome and debilitating autoimmune diseases—such as Crohn's disease and ulcerative colitis—that arise when the immune system is unable to regulate itself, leading to chronic inflammation and damage to the body's own tissue.

Some people have genetic profiles that make them more susceptible to ailments such as these inflammatory bowel diseases. But human genes explain only about 30 percent of the risk for Crohn's disease and 10 percent of the risk for ulcerative colitis, says Gary Wu, a gastroenterology researcher at the Perelman School of Medicine at the University of Pennsylvania. "That means most of the risk of developing these diseases is actually environmental." And some of these environmental forces could trigger harm-inducing changes in the microbiota, shifting the immune conditions to favor the development of an immune-mediated disease. For clues, he says, we can see that the rate at which people are getting Crohn's has risen rapidly in many countries over recent decades. "It used to be that in Asia—China, Japan, India—inflammatory bowel disease was relatively uncommon," he says. "But as those cultures become more industrialized, they're seeing a rapid rise in the incidence of these diseases." And he thinks shifts in the gut microbiota as a result of these industrialized lifestyles might be a contributing factor in the diseases' growing frequency.

One of the other potential curses of clean living is more macro: a lack of parasites. Although not having worms is generally viewed as a good thing—for us and for our pets—they were common companions for much of our evolution. Many scientists have noted it is likely that our bodies, evolutionarily speaking, *expect* them to be there. This is part of the "old friends" hypothesis and suggests that our immune systems are in a way missing them and acting out in their absence.

Along with so many of our other lifestyle changes (pasteurized foods, clean buildings, industrialized diet, annual antibiotics), banishing parasites (tapeworms, hookworms, pinworms) has paralleled the rise in allergies and autoimmune diseases in wealthy countries. It might seem outlandish to point to this shift when there are so many other dynamics at play in determining our health, but researchers have been studying what happens in patients who have worms *re*introduced. And the results have been pretty stunning, even if the process is not for the faint of heart—or weak of stomach.***

Give someone worms to help their asthma? This idea has turned out to be not as outlandish as it sounds. I had the opportunity to visit with researchers at Harvard Medical School in some of the earlier days of this testing, and at the time, the concept seemed rather abhorrent (though props to the researchers who convinced their institutional review boards). For these investigations, scientists typically use organisms that are not native to humans, such as pig whipworms, to reduce the odds of persistent infection or accidental spreading to others. The worm is administered orally in is egg form (usually in a drink—bottoms up!) and allowed to mature in the gut. There the worms, known as helminths, have a distinct effect on human genes involved in aspects of the immune system. The changes that come about while these worms are present can mitigate Crohn's disease, ulcerative colitis, and possibly even celiac disease and allergies. And the search for other improved conditions continues.

Newer research is examining the complex interplay not just between gut parasites and our own genes, but also between parasites and microbes. One study found that some worms might encourage the growth of bacteria that fight certain microbial strains that can cause inflammation. This sort of interplay might also help explain many of the correlations researchers have been seeing for years between helminth infections (whether natural or induced) and protection against illnesses such as Crohn's disease.

*** Really, the more appetizing stuff *is* coming later.

• • • • •

Our gut inhabitants also help the immune system by warding off infection. One way this is done is by training the immune system to make it more adept at fighting off infectious bugs, such as those that might cause a cold, flu, or gastroenteritis. Microbes don't always have to enlist the immune system, though. A gut filled with a healthy, diverse microbiota simply has less room for a pathogen to move in: the resident microbes are already using up most of the real estate—and food. In experiments giving mice the harmful *Salmonella* bacterium, researchers saw that in many cases, the local microbiota simply outcompeted the would-be invader, preventing the infection. Mice given antibiotics to wipe out most of their normal microbes before being infected, however, were much more likely to become ill. A new weed will have an easier time overtaking a freshly plowed dirt field than in a densely populated rain forest.

Pathogens don't have to come from the outside. They can also grow and develop from within. Any big disturbance of the established gut microbiota—say, with a course of antibiotics—can allow existing survivors to grow and take over. If you had a lingering colony of potentially pathogenic *Escherichia coli*, these few microbes now have more room and resources to spread and multiply. Research has shown that when the number of Bacteroidetes bacteria goes down, the susceptibility to at least one species of pathogenic bug rises. And in large numbers, even theoretically harmless microbes can cause illness.

The microbiota can also fend off invaders in the way that microbes have been defending their home turf for billions of years. As long as there have been microbes, they have been making compounds to keep other microbes at bay. It's been an arms race ever since, even spawning compounds to disarm those compounds (the original antibiotic resistance). We like to think we clever humans invented antibiotics, but it was actually the microbes. We just adapted them for our use.†††

Mood-Altering Microbes

In addition to being a crucial immune center, the gut also has deep connections to the nervous system—including 100 million dedicated neurons.‡‡‡ This partnership has researchers also searching the microbiome for connections to mood and to brain disorders. One tantalizing clue is that about 80 percent of the body's serotonin—a deficiency of which is linked to depression—is made in the gut. And researchers have indeed found that people with major depression have different gut microbe patterns than healthy individuals do.

In addition to being connected to the many millions of neurons, the colon and the brain are also directly connected by the powerful vagus nerve. This provides another means of influence on the central nervous system from the microbiota. Animal studies have linked this pathway not only to the control of inflammation but also to brain function and mood. Early research also shows that mice raised without gut microbes develop differently, neurologically. Some scientists are even looking for connections between the microbiome and autism—another disease that is on a steep rise.

††† If you remember some of your scientific history, it was mold (specifically, the fungus now known as *Penicillium chrysogenum*) growing in one of Alexander Fleming's petri dishes in the 1920s that was found to have kept his harmful bacterial cultures at bay. From this discovery, the drug penicillin was eventually developed, successfully treating the first patient in 1942.

‡‡‡ This might not sound like a lot, especially compared to the 86 billion neurons or so in the human brain. But 100 million neurons is still a good 10 million more than your average hamster has in its entire nervous system. So maybe it's time we trust our guts a little bit more. They have a lot more going on than we previously realized.

Patterns in the microbiota also seem to be linked to anxiety. Mice induced to symptoms of anxiety and depression by frequent early separation from their mothers have a different gut microbiota than control mice who were allowed to spend a happy puphood alongside their moms. Adding yet more intrigue to this finding, researchers discovered that when microbe-free mice are separated following the same protocols, those negative behaviors actually fail to develop. This suggests that microbes play a role in the development of mood and behavior—and can be shaped from an early age. To this end, scientists have found that transferring the gut microbiota from people suffering from anxiety into germ-free mice also transferred anxious behaviors to the mouse subjects.

We might also be able to look outside the body for microbe-mood interactions. Nonresident microbes seem to be able to switch off murine depression. How can we tell? It's difficult to have a heart-to-heart with a mouse, so scientists have developed other ways of figuring out how mice are feeling. One of those is a swim test. Mice are placed in a small tank of water and tested to see how long they keep paddling—absent an escape route—before they give up (and are rescued by researchers). Mice showing depressive symptoms will give up much more quickly than healthy control mice. A research team at University College Cork in Ireland decided to see how they could start to alleviate these symptoms in depressive mice—and help them swim longer. In one study, they tested a commercially available antidepressant versus a strain of *Bifidobacterium*. Both worked. As did using a strain of *Lactobacillus* in another trial.

In other research, Actinobacteria (such as *Mycobaterium vaccae*, which is found in the soil and which we can come into contact with through water or plants) and Firmicutes (especially lactobacilli) seemed to reduce anxiety in animals.

Some studies are even finding effects in humans. One such

examination scanned the brains of healthy adult volunteers using functional magnetic resonance imaging (fMRI). A subset of the participants then drank probiotic milk beverages (inoculated with *Bifidobacterium animalis, Streptococcus thermophilus, Lactobacillus bulgaricus,* and *Lactococcus lactis*[§§§]) daily for a month. After the month was up, control participants' brains looked the same, but those who had been drinking the probiotic blend showed changes in activity in the areas of the brain related to emotion and sensation.

We have a ways to go before specific microbes can be prescribed for mood disorders, but this research is laying intriguing foundations. And these findings are leading researchers to the investigation of a new class of probiotics known as psychobiotics.

• • • • •

Despite all of this advanced scientific research, so far the best evidence of the microbiota affecting health comes from a rather crude-sounding procedure. It involves transferring fecal matter from one person to another. Although leeches or bloodletting might seem slightly more appealing (and even perhaps more hygienic), for a specific population of patients receiving the fecal transplant, the caca calculation has been completed. And the benefits outweigh the revulsion.

These patients suffer from a nasty and often very dangerous intestinal infection known as C. diff, short for *Clostridium difficile*.[¶¶¶] The infection—an overgrowth of this harmful bacterium in the intestinal

---

§§§ The three former of which are often found in fermented dairy products, *Streptococcus thermophilus* and *Lactobacillus bulgaricus* being the active bacteria in yogurt fermentation and *Lactococcus lactis* being found in some aged cheeses, such as Gruyère.

¶¶¶ The difficile (French for difficult) in the bacterium's name, for those suffering an infection, might seem a bit of an understatement.

tract—can cause diarrhea, fever, severe stomach pains, and in some cases even death. Patients often acquire a C. diff infection in the hospital, although cases originating outside hospital walls are on a worrisome rise. Many of those who get C. diff infections fall ill after rounds of broad-spectrum antibiotics, which had wiped out other bacterial species (some bad ones, to be sure, but also many good and neutral ones). This clears a path for the pathogen to move in and multiply. It creates compounds that break apart the bindings that keep the gut lining together, as well as other toxins that then escape and circulate throughout the body. Many cases of C. diff can be treated successfully with yet more rounds of antibiotics. But more and more frequently, this approach has been failing. And patients are left, several rounds of antibiotics later, with worsening infections and grim prognoses. So doctors have turned to fecal microbial transplants, or FMTs for short.

There is actually a long history of taking this approach for severe gastrointestinal illness. Even before the infective bacterium was connected to the condition, a rather brave (or crazy) group of Denver, Colorado, physicians working together in the 1950s began treating patients suffering from severe intestinal symptoms with fecal enemas—with a surprisingly high success rate. Long before that, sixteenth-century Chinese physician Li Shizhen recommended a feces-based "yellow soup" as a treatment for abdominal pain, diarrhea, and other gut ills. Even earlier, in the 4th century, Chinese writer and alchemist Ge Hong prescribed an oral dose of fecal suspension for severe diarrhea or food poisoning. "Reseeding" is apparently not such a new idea after all.

Fecal transplants, which introduce microbes from a healthy person's digestive tract into an ill person's, are quite the broad brush in an era of personalized precision medicine. But to most everyone's surprise, they were so successful in rapidly vanquishing *Clostridium difficile* that early trials to test their efficacy were halted, and participants who were being treated with more antibiotics were given the transplants instead.

With a 90 percent cure rate, withholding the transplant treatment from any sick participants would have been unethical.

Researchers have now refined the process (which began by installing the transplant via enema or infusion through a nasal feeding tube) and have created pills from screened fecal donors (yes, essentially poop pills). Another approach is autotransplantation. That is, if a patient is to undergo substantial antibiotic treatment, he or she could "bank" their microbiota (in the form of frozen stool). And after the procedure, if their microbiota did take a big hit, they could then restore their own unique community. Start-up companies are also in the process of attempting to build artificial microbiomes from pure lab-grown microbe communities, eliminating at least some of the yuck factor. However, creating a successful, vibrant community from the ground up has proven challenging.

With the tremendous success treating C. diff, scientists are now looking at a vast array of other conditions that might be treatable via a new—or at least remodeled—microbiome. One obesity study showed that infusing the intestinal microbiota from lean donors into people with metabolic syndrome helped to resensitize recipients to insulin (a move away from developing diabetes). Other teams are looking to use it to treat inflammatory bowel disease and even autism.

Manipulating our microbes to control our health might sound futuristic, but it has deep roots in thousands of years of traditional healing and traditional eating, which were often pretty much one in the same. We are finally waking up to this stored wisdom, and diet is returning to its rightful focal point as a key for holistic health. Thankfully, diet is an accessible—and enjoyable—form of preventive health management. So let's look to where these lessons come from—and to generations of cooking and eating, harnessing the power of the microbe.

CHAPTER THREE

Feeding the Microbiome

Throughout our lives, the microbes in our bodies come not just from birthing and infancy, but also from the environment—from water, air, buildings, pets, colleagues, our partners, and what we eat.* Our diets have the power to influence what microorganisms we introduce to our bodies—and how hospitable our bodies are to beneficial organisms already there. It is also one of the things we have the most control over. And it has the potential to bring pleasure and discovery in the process. In this realm, there are plenty of things that most of us could do better when it comes to nourishing a diverse and thriving microbiome.

Feeding your existing microbes might not sound quite as exciting as introducing new and exotic ones by the billions. But many scientists argue that to truly do the best that you can for you and for your microbes, ensuring that they have the right food is essential.

* Beyond food that contains live organisms, most natural food has its own microbial landscape—even if we wash it.

Be a Good Host

So what is the best way to feed our microbes? In one word: fiber. We have long known that fiber is good for us. It helps reduce caloric intake and maintain regularity. But it is also perhaps the most powerful tool we have to help our native microbes. It is their bread and butter, so to speak.

Fiber is made up of long chains of carbohydrates.[†] Because these carbohydrates are connected by complicated bonds, these molecules are difficult—and sometimes impossible—for us to digest. We humans simply don't have the enzymes necessary to break down many types of fiber. And that means these compounds end up intact down in the lower intestine, where helpful microbes can feast on these cast-offs. When these compounds encourage the growth and health of beneficial microbes, they are known as *prebiotic*.

In recent years and decades, we haven't been very good at providing this expected fodder to our native microbes. And without fiber to nourish them, their populations take a dive, leaving us without their many benefits.

The average American now consumes about 15 grams of fiber per day, roughly half of what the U.S. government recommends. Those 30 or so recommended grams of fiber itself is likely about a third (or less) of what a more traditional diet might offer. Even the high end of this range is a fraction of what our ancestors probably ate every day. All of that means that we're eating just 10 to 15 percent of the fiber that our microbes would have expected. And they seem to be feeling the deficiency—as are we.

"Humans used to eat maybe 100 to 150 grams of fiber every day," says University of Nebraska–Lincoln food scientist Robert Hutkins.

---

† Thus they behave very differently in the body than foods we typically think of as "carbs"—pasta, white bread, and the other simple carbohydrates we encounter so often these days.

"And that certainly had an influence over our microbiota for tens of thousands of years. It's only in the last fifty to one hundred years that we've completely done a one-eighty by eating pop and chips. Our gut microbiota would've been completely different by seeing that much naturally occurring prebiotic fiber."

For example, one archeological study of cave sites in the Chihuahuan Desert inhabited by humans for some 10,000 years found evidence of "intensive utilization" of local plants high in prebiotic fibers. Clues gathered from cooking materials, human skeletons, and coprolites (fossilized excrement) suggest that the inhabitants were eating some 135 grams of a specific type of microbe-feeding fiber (inulin) each day. As we will see later, this important prebiotic fiber feeds microbes that provide us with services like creating anti-inflammatory compounds. Estimates place current U.S. and European consumption of this form of fiber, on the other hand, at just "several grams" per day. Rather a far cry from the earlier culture. And one that our microbes don't seem to be happy about.

The ancient desert dwellers might have been an exceptional case, but we know that through history, as a rule, humans had much more fibrous meals. Study after study points to the diversity of Paleolithic diets. An investigation of a 23,000-year-old site in Israel uncovered that the local cuisine included more than 142 different species of plants (including seeds, nuts, fruits, and cereals).[‡] Although the work didn't specifically investigate the fiber content of the residents' diet, the impressive diversity of plants at the site suggests meals rich in fiber—and many different forms of it at that.

Even much more recently, humans were consuming a wide range of fibers regularly. A man known now as Ötzi lived near what is now the Austria-Italy border around 3285 BCE and was frozen in a glacier, only

‡ This all before the advent of global food supplies. Oh, and of agriculture. With the current availability in mind, how many plant species have you eaten today?

to be exhumed in the 1990s. A subset of researchers (the stomach team) examined the contents of his digestive tract. They found that before his death, at around age forty-five, he had recently eaten a wide variety of foods, including wheat bran, barley, flaxseeds, local fruits, legumes, as well as roots and goat and deer meat. Beyond his last meals, they also discovered that he also harbored a high diversity of microbes in his large bowel.

Some fiber-rich diets have not totally died out. Studies on fiber intake in the mid-twentieth century found that many people on the African continent still eating relatively traditional diets were consuming 60 to 140 grams of fiber per day.

Recent studies have shown that Africans who eat traditional diets have gut populations dominated by *Prevotella* bacteria (indicative of a carbohydrate—and fiber—heavy diet). In contrast, African Americans consuming a standard U.S. diet tend to have a preponderance of *Bacteroides* (which are associated with a diet heavy in animal products and more common in the United States). Although it is only a correlation, African Americans are also often at a higher risk for colorectal cancer. But researchers found they could rapidly decrease one of the markers for this risk in those who went on a high-fiber diet (more than 50 grams per day) for just two weeks. The researchers saw the opposite shift in African participants when the study participants consumed a high-fat, high-protein, low-fiber diet.[§] Both of these dietary changes altered the microbiota as different foods for microbes waxed and waned.

This is shift is strikingly apparent in studies of people who have migrated to wealthy nations and are suddenly eating a drastically different diet. Not only do their microbes change, but so, too, do their risks for many Western-associated diseases, such as inflammatory bowel disease (IBD).

§ They also saw a shift in microbes related to inflammation and IBD, such as *Bilophila.*

In a sense, most all of us have experienced this, if on a longer time scale. Even if our diets have stayed the same for most of our life, they have certainly changed over recent generations.

When our diets change and stay altered for years, what does that mean for the microbiota? Erica Sonnenburg, a senior research scientist at the Stanford University School of Medicine, is trying to figure that out. "We started doing these experiments, looking at diet and the microbiota," she says over a lunch of legume-heavy salads at a Stanford cafeteria with her husband and fellow microbe researcher, Justin Sonnenburg. "When we put mice with a human microbiota on a diet where we've removed all the dietary fibers—so basically what these microbes would rely on—we'd see a pretty quick crash in the numbers of types of microbes that we see in the gut," she says. "If we do this on a very short time scale, a few days, we see it pretty quick—within a day. Then when we introduce the dietary fiber, everything seems to pretty much go back."¶ And that's a good thing, as it means that the microbiota can be rescued just by adding fiber back into the diet.

"Then we started wondering," she says, "it's not really what's happened in the [Western] diet. Really, what we've done is, we've removed dietary fiber from our diets, and we've done it for long periods of time, during which we've had children and put them on low-fiber diets. And so the question is: What happens over longer periods of time in these mice? And could it happen in humans as well?

"What we've found," she says, "is that even within one generation, if these mice are on a low-fiber diet for an extended period of time—several weeks—when we reintroduce dietary fiber, there is some regaining of that microbial diversity. But it doesn't appear to be complete."

¶ Justin Sonnenburg chimes in that they had joked about titling the resulting paper "It's Not Too Late—Eat Your Dietary Fiber!"

That suggested to them that long-term fiber changes could potentially alter the gut microbiota permanently.

To test this idea further, the Sonnenburgs, who also cowrote the book *The Good Gut*, and their colleagues allowed mice with human gut microbiota to reproduce while on either regular high-fiber mouse chow or altered low-fiber chow. The big question was: If mothers were giving birth to pups while they themselves had an altered microbiota, what did that mean for future generations? "You can imagine a scenario where, for certain types of microbes, there would be no transfer because it's so low-abundance that it's just not getting transferred," Erica Sonnenburg explains. "What we see is there's a major drop-off from generation one to generation two. Even another drop-off from generation two to three, then by three to four, it's pretty much that the microbiota has achieved some stable state of low-fiber, low diversity."

That's a pretty grim tale. Especially if we consider the particularly precipitous fiber nosedive that Western diets have been on for the past few generations. So the Sonnenburgs wondered, could microbes lost over generations of low-fiber diets be restored by adding fiber back into the diet? "In the last generation that we had—generation four—we reintroduced dietary fiber," Erica Sonnenburg says. But restoration of gut microbiota didn't occur. "They just cannot regain that diversity. So we think that with the successive passaging these microbes just aren't there, they're not in the gut anymore." Even if certain populations of microbes were persisting at low levels in earlier generations, "it's so low that it effectively is not there," she says. Even with fiber back in the diet, "they just can't get back to where they were."

So what does all of this mean for us? "That has massive implications for what we've done in the Western world to our microbiota," she says. These worrisome impacts, as we will see, are many.

The fibrous complex carbohydrates that specifically feed our beneficial microbes have many microbe-mediated benefits. Researchers have found, for example, that prebiotics might help our microbes keep pathogens at bay, improve the immune system, boost absorption of minerals, encourage weight loss and satiety, reduce diarrhea and allergies, decrease inflammation, reduce IBD symptoms, increase insulin sensitivity, protect against colon cancer, and possibly reduce some risks for cardiovascular disease. Even if all fiber-filled foods resembled the old cardboard-like fortified cereals of the 1980s, all of these benefits alone might still might be reason enough to eat them. But fortunately for us, beneficial fibers are easily available in a wide range of enticing foods from traditions across the globe. We will dip into some of these dishes later on.

In the meantime, we are still learning more about these crucial compounds that feed our native microbes. Among the best studied of them are inulin, fructooligosaccharides, galactooligosaccharides, and resistant starch. Where are these mouthful-of-a-name compounds found? Many of them occur naturally in an astonishing variety of plants—tens of thousands of which are edible. But you don't need to travel across the globe or take up foraging. Many of them are found in the standard stock of local grocers.** For starters, you needn't look further than bananas and leeks.

• • • • •

Perhaps the best-studied type of prebiotic is inulin, which we encountered in great quantities in the ancient Chihuahuan Desert diet. Inulin is a polysaccharide (a long chain of smaller carbohydrate molecules, held together by anywhere from a few to dozens of bonds) and offers many of the benefits mentioned above by feeding our

** As the Sonnenburgs note, "The entire produce section of the grocery store should have signs and stickers: Contains Prebiotics!"

beneficial native microbes. It occurs in a wide range of fruits and vegetables, but its highest concentrations are found in chicory root (from which it is isolated for supplements and food additives). It is also present in leeks, onions, garlic, burdock, asparagus, and underripe bananas, among others—as many as 36,000 plant species in total.

One of the best dietary sources of inulin is an unassuming tuber: the sunchoke, also known as a Jerusalem artichoke. These humble root vegetables look a bit like extra-knotty gingerroots, a golden-brown skin covering segments joined at unpredictable angles. These vigorous, potato-like tubers have their devotees—and their detractors. Recently they have been praised as a worthy, low-glycemic index food, high in complex carbohydrates and low in simple starches. In earlier decades, however, they were cast off as starvation rations, animal feed, and in the 1980s, the center of a biofuels pyramid scheme. But for us (and more important, for our microbes), they are a rich source of this prebiotic compound. They also have a pleasant, subtly sweet flavor, making them excellent roasted, mashed, or even sliced raw into a salad.

Even without consideration for the gut microbiota, inulin has also long been added (sometimes listed on products as chicory root extract) to foods as a low-calorie replacement for additional fat or sugar. As one team of researchers describes, these types of inulin "have a neutral, clean flavor and are used to improve the mouthfeel, stability and acceptability of low-fat foods." Now inulin is also being packed in as a "functional" food ingredient in its prebiotic capacity.

One key way inulin helps us out is by is by encouraging the growth of bifidobacteria, lactobacilli, and other favorable gut microbes. One study found that eating extra inulin (10 grams per day for a month) reduced participants' traveler's diarrhea. And another study found that 8 grams of inulin per day helped teens absorb more calcium. Too much of it, though—especially all at once and for those who are not used to it—can cause gas and bloating (by-products of microbial fermentation).[††]

• • • • •

Another key prebiotic is oligofructose or fructooligosaccharides (FOS), which are chains of fructose molecules. These are not as long as some other prebiotics (containing up to just ten bonds) and are thus often broken down in the earlier areas of the large intestine. For commercial extracts, FOS is often made from agave plants but can also be found in inulin-containing plants (sunchokes, onions, leeks, etc.) as well as in grains such as barley and wheat. The all-star chicory root is some 15 to 20 percent inulin by weight and 5 to 10 percent oligofructose. This compound is often added to foods as a sweetener (it's about a third to half as sweet as sugar). Like inulin, it has been shown to stimulate the growth of bifidobacteria species. A study of FOS and inulin supplementation found that adding them to participants' diets seemed to reduce inflammation markers and helped patients with IBD. And another study found that increasing the intake of this compound might also help reduce weight.

Galactooligosaccharides (GOS) are made up of chains of galactose, which, like FOS, are often at least partially broken down by the time they reach the lower large intestine. These compounds occur naturally in low concentrations in milk (from cows as well as goats, sheep, etc.) and in slightly higher concentrations in yogurt. It is similar in structure to compounds found in breast milk, known as human milk oligosaccharides.

---

†† "You can put a couple spoonfuls in your coffee—see what happens," says David Mills with a knowing smile. "I know what happened to me. I blew up like a balloon." But the body and microbes can be gradually accustomed to increased amounts over time.

But those added to foods (such as to infant formulas) are generally manufactured (often in part by the miso-making fungus *Aspergillus oryzae* as it feeds on lactose). One study discovered that GOS supplementation increased bifidobacteria and improved symptoms for individuals with IBS. And another study found that 5.5 grams per day of GOS also reduced traveler's diarrhea.

Among all of the complex, unwieldy categories is a surprise entry to the prebiotic cannon: starches found in many foods, including some simple carbohydrate foods such as potatoes, white rice, and even pasta. There is just one catch: these foods must first be cooked, then cooled.

Many types of resistant starch are found in a wide variety of foods. Lentils, corn, and kidney beans have the highest amounts, but it is also prevalent in barley, black-eyed peas, rice, wheat, corn flour, oats, and several other cereal- and grain-based products. It can also be found in underripe bananas and mangoes. Generally speaking, more processing—say, taking wheat from whole grain to flour—means less resistant starch.

Then there is so-called retrograded resistant starch. This form occurs when the cooked simple starch—which has incorporated water and turned into a gel-like substance—cools and crystallizes, making it impenetrable to our digestive enzymes. This means it passes more or less untouched down to our hungry microbes, where they can have at it. Like other prebiotics, resistant starches can help increase stool bulking and provide other intestinal benefits.

• • • • •

With our ever more processed diets, we have to be extra vigilant to incorporate these once-common compounds in our meals. To ensure that we get these motley prebiotics, nutritionists and microbial researchers alike suggest we eat a wide array of fibers. Because these prebiotic fibers come in different lengths and complexities, they will be broken down at different points in the digestive process. For example, FOS is relatively short, so it is fermented by bacteria relatively quickly, higher up in the intestine, whereas inulin and resistant starches are larger and take more time for microbes to ferment, providing foods for microbes farther down. Simply increasing one specific type of fiber can actually reduce microbial diversity in the gut because the microbes that are extra efficient at using it will eventually crowd others out. Ironically, despite the unprecedented abundance of food options available to us today, we often choose to follow extreme exclusionary diets (even in the absence of diagnosed allergies), cutting out foods or entire groups of foods for the promise of quicker weight loss or clearer thinking. Such drastic elimination of foods and categories from the diet can actually make it hard to reintroduce those foods later, in part due to changes in our microbiome, which can lose the ability to break down certain compounds. Variety is more than the spice of life—it is also the key to good gut health.

• • • • •

When fibers are scarce for many of our beneficial microbes, our residents do have one trick up their sleeves. They can start eating *us*—that complex carbohydrate gut lining that many microbes can turn to in times of famine. Unfortunately, this carb coating is also the crucial protective mucus layer that separates our microbes from the delicate gut wall—and the bloodstream and the inner body beyond.

The mucus—composed mostly of mucin—is there in part to sustain our good microbes. Mucin is a natural compound that our gut makes to keep the beneficial bugs around through thick and thin, so to speak. And we renew it constantly. But in those prolonged thin times, microbes might turn too eagerly to this food source. And as we saw earlier, a broken gut barrier can lead to inflammation (through the "leaky gut" condition in which microbes and food particles escape from the gut into the bloodstream‡‡). This condition can, itself, select for inflammation-causing microbes, launching the body and microbiome into that harmful cycle of inflammation.

As we have seen, eating a diet high in prebiotics can improve gut barrier function. A high-fat diet, on the other hand, has been shown (at least in animal studies) to reduce the thickness of the mucus lining—which might be one of the ways a fatty diet leads to persistent inflammation and contributes to metabolic diseases and other related conditions. This is visible in the case of obesity. Studies have found that obesity is correlated not just with a reduction in microbial diversity, but also with a reduced thickness of the gut's protective mucus. So, when in doubt, choose the fiber.

Scientists have delved even further into how crucial fiber is to the integrity of the gut lining. Eric Martens, a microbiologist at the University of Michigan, and his colleagues wanted to see just how intricately tied fiber intake is to mucus layer thickness. They started with germ-free mice and gave them a selection of human gut microbes—some of which are known to be able to feed on mucin. One group of mice was fed a high-fiber diet, another was fed a fiber-free diet, and a third group alternated daily from high fiber to no fiber—"like what we would do if we were being bad and eating McDonald's one day and our whole grains the next," Martens explains. In the high-fiber group, the mucus layer was relatively thick. In the no-fiber group, "the mucus layer becomes dramatically diminished," he says. But even "when you oscillate these diets on a day-by-day basis, you get something in between—which tells us that even eating your whole-fiber foods every other day is still not enough to fully protect you from the bacteria that live in your gut," he says. "You need to eat a high-fiber diet every day to keep a healthy gut."

‡‡ Some bacteria can also release endotoxins into the body, which have been linked not just with inflammation but also with obesity and insulin resistance.

• • • • •

When we do eat food that contains prebiotic fibers, gut microbes in turn repay us by making compounds that can help quell inflammation or defend us against infection. These compounds, known as metabolites, are microbial by-products, expelled during microbes' metabolic process of digesting food that comes their way. Fortunately, these by-products just happen to be beneficial for us.[§§]

Many of these compounds are types of short-chain fatty acids (SCFAs). Although they might not sound like things you would want in your gut, they actually perform a range of services that are crucial for our health in the lower intestine and beyond. For one, these water-soluble molecules are easily absorbed into the bloodstream for transport and use around the body, providing essential energy for cells in the body—from colon cells to brain cells.

As their name indicates, these compounds are acidic, helpfully lowering the pH of the intestine. This makes the gut extra-inviting for beneficial bacteria, such as *Lactobacillus* and *Bifidobacterium* species, which thrive in acidic environments—and less hospitable for pathogenic microbes. The helpful bacteria in turn can produce more SCFAs. These compounds also help to regulate water and sodium in the gut and encourage absorption of crucial minerals such as calcium. Among the top three SCFAs are acetate, propionate, and butyrate.[¶¶]

Acetate in particular feeds muscles, the brain, and other tissues. Propionate is taken to the liver, where it might help to lower cholesterol

---

§§ Some by-products, of course, are not quite so pleasant, such as methane—an occasional "off-gas"—along with carbon dioxide and hydrogen, which come from gut bacteria at work. But as David Mills mentioned before, the body generally adjusts.

¶¶ These compounds might sound familiar, perhaps from the world of fermented foods. That's because a lot of similar types of bacteria that ferment foods outside of the gut are also busy fermenting foods inside the gut.

levels. It, along with butyrate, also regulates immune and intestinal function. Butyrate is a preferred food of the cells that line our large intestine. It helps these crucial cells grow and proliferate normally, making it "the most beneficial, in terms of colonic health," according to one paper on the subject. The resistant starch we learned about earlier seems to be particularly adept at being metabolized into butyrate (via our microbes)—another reason it's such an important prebiotic for gut health. Butyrate has also been used as a general measure for the health of the microbiota—and for health overall—because unhealthy microbiota put off less of this compound. For example, having low amounts of butyrate has been linked to type 2 diabetes. Furthermore, a reduction in butyrate-producing microbes also seems to occur in patients with colorectal cancer.

Beyond these health links, early studies in animals show another reason to feed microbes the foods they need to make these fatty acids: protection from food allergies. As we saw earlier, a lack of prebiotics in our diets might lead to leaky gut, which could allow food particles to escape the intestine and enter the bloodstream, causing the immune system to rev up. Researchers have also traced another potential path for food allergies back to the microbiome. They found that mice prone to developing peanut allergies who were fed a high-fiber diet, which fostered a robust SCFA-producing microbiota, were protected from getting the food allergy. The same experimental mice on a more Western high-fat, high-sugar, low-fiber diet were much more likely to have an allergic reaction to peanuts. What was happening? It turns out that the fatty acids manufactured by these well-fed microbes actually bind to certain immune cells that help to calm the immune and inflammatory response, offering protection from the potential food allergy. Just to be sure it was the microbes doing the work, the researchers then transferred only the microbiota from the fiber-eating, allergy-protected mice into a new group of germ-free mice. Sure enough, these other mice were less likely to have an allergic reaction to peanuts. So perhaps to help prevent some of these food allergies from developing in the first place, we can start by offering our microbes better fare.

Adding in more prebiotic food also means, simply, more helpful microbes. Some research has suggested that for every 10 grams of prebiotic carbohydrates that reach the gut microbiota, about 3 grams of additional bacteria blossom into life. That's roughly 3 trillion new organisms, just as a result of adding those 10 grams of microbe fodder each day. Not a bad trade-off for eating some extra whole grains—and cold potato salad.

So when it comes to prebiotics, instead of "build it, and they will come," think "eat it, and they will multiply"—and possibly even help protect you from a range of increasingly common health concerns. All it takes is paying a little more attention to what you're feeding your microbes.

• • • • •

This pattern has been borne out in study after study, year after year. One striking example comes from an investigation of two groups of children—one living in Italy and the other in Burkina Faso. The researchers studied the diets and lifestyles of the children, who ranged in age from one to six years old. The children in Italy lived in the city of Florence. There they ate a pretty standard Western diet, which was high in animal protein, fat, sugar, and simple starches. It was also predictably low in fiber, with younger children eating about five and a half grams of fiber per day and older children consuming about eight and a half grams per day. The children in the West African nation came from the Mossi ethnic group and lived in huts in a small rural village. Their lifestyle, the researchers noted, was likely fairly representative of the typical human existence after the Neolithic agricultural revolution, about 10,000 years ago. They ate little animal protein (occasionally some chicken, or termites during the rainy season), but lots of fiber and complex starches. Their staples included millet and sorghum (eaten in the form of a porridge), black-eyed peas, and vegetables. All of this amounted to nearly double the amount of daily fiber the European children ate, with younger children consuming about 10 grams of fiber per day and older children getting a little more than 14 grams. This might not quite be at the ancestral adult hunter-gatherer levels, but a doubling of fiber intake—especially while children and their immune and nervous systems are still maturing—is a big boost for the microbes and for future health prospects.

When the research team looked more closely at the bacteria living in the two groups' guts, they found distinct differences. The Italian children possessed the expected Western pattern, high in Firmicutes (about 64 percent) and low in Bacteroidetes (about 22 percent). The Burkina Fasan microbiome had the reverse ratio, with abundant Bacteroidetes (about 58 percent) and many fewer Firmicutes (about 27 percent). What is even more interesting is the presence and absence of certain bacteria from each group. The Mossi children possessed *Prevotella, Treponema,* and *Xylanibacter* species—none of which were found in the European children. These bacteria have loads of genes to efficiently break down tough prebiotic compounds found in a plant-heavy diet. European children, on the other hand, were much more likely to have higher concentrations of harmful bacteria.

Another telling observation about the microbiome's power is in the number of calories the children ate. The researchers were careful to select all healthy children who had grown about equally for their age. But despite being of similar weight, the children in Burkina Faso were consuming only about two-thirds as many calories per day as the Italian group. "Diet plays a central role in shaping the microbiota," the authors note. And "the presence of these three genera could be a consequence of high fiber intake, maximizing metabolic energy extraction from ingested plant polysaccharides." So thanks to their *Prevotella*, *Treponema*, and *Xylanibacter* bacteria, the children from Burkina Faso were able to get more out of their mainly plant-based food, requiring them to eat fewer calories each day.

In addition, the Burkina Fasan children also had substantially more short-chain fatty acids in their fecal samples, including almost four times as much beneficial butyric and propionic acids as the Italian children. This likely comes from having the right microbes for the job as well as eating a high-fiber diet. As the authors note, "whole grains are concentrated sources of dietary fiber, resistant starch, and oligosaccharides, as well as carbohydrates that escape digestion in the small intestine and are fermented in the gut, producing short-chain fatty acids."

Beyond the microbe-driven health benefits for the Mossi, the study also suggests lurking danger for those living a Western lifestyle. "The microbial simplification harbors the risk of depriving our microbial gene pool of potentially useful environmental gene reservoirs that allow adaptation," the authors note. "The lessons learned," they continue, "prove the importance of sampling and preserving microbial biodiversity from regions where the effects of globalization on diet are less profound." Not only are we losing these important microbes individually, but we are also on the cusp of losing many of them globally. And possibly for good.

The findings are not, of course, to say that we should all return to eating termites in the rainy season. But they are a stark reminder of just how much we have shifted our diets and lifestyles—and the impacts those shifts are having in the invisible landscape of our guts. And we are just beginning to sort out all of the long-lasting health implications of these changes.

Not all of the changes that have taken place in the recent centuries have been bad, of course. One of the major public health successes of the twentieth century was to rid our food supply of most all of the pathogens that lurked there. This has been a terrific victory for saving us from acute—and sometimes deadly—foodborne illnesses arising from tainted milk or botulism-filled meats. But heating, bleaching, and otherwise processing foods has also vanquished many of the beneficial and innocuous microbes that humans had been ingesting with every bite for millennia.

Beneficial Visitors

Now that we've learned all about our native microbes and the food they like to eat, it's time to explore the enticing, mysterious world of probiotics—and the bubbling, distinctive fermented foods where we can find them. These are the microscopic denizens of kimchi, kefir, and kombucha. They are the microbes we seek out to bring healthful benefits as they pass through our digestive tract. And we cultivate them through the process of fermentation.

Fermentation as a method of food transformation is seeing a rousing resurgence. Before the invention of safe canning and before the advent of widespread refrigeration, food could be preserved by being dried, salted, pickled in vinegar, or fermented. It was often a matter of ensuring adequate calories and nutrients through lean times. These days, of course, fermenting is not just a way to keep food around a little longer. It is also a way to bring new flavors, textures, and interest into dishes. And of course it comes with a promise of health.

But before we go any further, there is one important point to make: Not all fermented products are technically probiotic. Sorry, but it's true. For starters, just because that sourdough bread was created from a fantastically rich live culture (known as a symbiotic colony of bacteria and yeast, or SCOBY) and that beer was made with yeasts, doesn't mean that these microbes have been scientifically proven to be beneficial or that they make it into the finished, distributed product alive—let alone into your intestines intact. What we are primarily interested in, for the gut microbiome, is microbes that *do* survive the journey through preparation, processing, and digesting.*** They must also be effective. Technically speaking, to be considered probiotic, a microbe must provide a measurable benefit to us humans.

Some probiotics have a good track record of helping out with specific conditions, such as antibiotic-associated diarrhea. Others might confer benefits for a variety of conditions. But not every strain can help with every condition. And that can lead to disillusionment. "Probiotics have become almost the embarrassing relative" of traditional medicine, says University College Cork microbiologist Colin Hill. "Probiotics were overhyped, overmarketed." Which has led to a divergence of feelings about them. "Consumer view now is that probiotics either work for everything or they don't work at all." But that is changing—if slowly, he says. And more people are beginning to pick up on the nuances of probiotic action.

Part of the problem is the way we talk about probiotics. It would be

*** There is, however, a spectrum of life for bacteria. Some are vivacious and can easily reproduce (counted in the commercial food and supplement worlds as colony-forming units, or CFUs). Others might well be alive but are not vigorous enough to be able to multiply. Then there are those that seem dead, but it might also be that we cannot find the right environment to revive them, as microbiologist Colin Hill at Ireland's University College Cork explains. So he at least shies away from stark classifications—"life is a dodgy enough term," he says. To make matters thornier, some scientists suggest that microbes don't even have to be alive at all to have an impact on us. In theory, simply introducing the microbe's shell (which is covered in a particular pattern of proteins) could activate certain signals from other microbes or from the immune system. As Eric Martens explains, as bacterial cells pass through, whether dead or alive, they might just "sort of tip the balance on the host immune compartment side toward potential benefits—especially farther up in the GI tract . . . Even consuming a lot of dead, say, lactobacilli that are in things like yogurt, could be beneficial if those *Lactobacillus* cell wall products hit receptors and turn on beneficial immunological pathways versus inflammatory ones." So, he says, "there could be benefit to even eating dead bacteria." Indeed, research done by Nestlé showed that one of their proprietary strains was effective in stimulating the immune system even when it was dead.

ridiculous to say "pills cure cancer" and "pills can fix your headache." You would probably want to know exactly which pill that was, how much you should take, and how often you should take it. Perhaps you could take different pills to treat the same condition—or the same pill to treat different conditions—but you generally wouldn't want to take ibuprofen for your cancer or chemo pills for your headache. However, consumers and even physicians have treated probiotics just that way, without much regard for species, strain, or dosage. Although most probiotics don't carry the same risk of many medications, which could be taken for the wrong indication or in the wrong quantity, they do behave very differently from one another. That gets back to the taxonomic distinctions. A herd of wild boars and a horde of pet hamsters are going to have very different impacts on the same environment (say, your living room). And with bacteria alone encompassing an entire *domain* of life, we are talking about the potential behavioral difference between a philodendron and a puma. And we haven't even included the fungi—and, if we want to go macro, parasites.

"I believe probiotic health effects are real," Hill says. We might just need to shift our thinking about them. So it is time to stop holding up the almighty singular probiotic as a cure-all—and time to start thinking of them like the varied individual organisms that they are. To be sure, much more research remains to be done to tease out the multitude of microbe species (and strains) and their particular impacts in particular cases.

The More, the Merrier

There are a number of ways we can intentionally consume these beneficial transient microbes. There are isolated discrete strains or blends in supplement form. There are also foods that have been inoculated with beneficial microbes, such as probiotic yogurts. And then there are those foods that have been allowed to ferment for the sake of making the food last longer, taste better, or become more nutritious. These are where our guts can find the most varied microbe palette—and sometimes beneficial prebiotics as well.

Manufactured microbe-filled products are not as diverse as traditionally fermented ones. Even in a well-stocked health food store, there might be only a short list of different species. Many "probiotic" yogurts have a couple of strains added in (though some might have a dozen or more). But what's a couple of groups—or even a dozen—in the crowded metropolis of our guts?

In naturally fermented food, however, there is often a far richer world of bacteria and fungi. That doesn't mean they're more likely to take up residence in the gut or that they are more likely to be useful. But it does mean that there are more opportunities for different—and potentially positive—impacts on us.

"I'm of the firm belief that all fermented foods—with their microbial ecology—are good for you in the sense that we should be consistently educating our immune system with a range of beneficial bugs," says David Mills, the University of California, Davis, microbiologist who studies microbes, food, and health. Microbes should continue to be a part of our diets. "I am a big fan of eating a diverse range of good bugs," he says. "Your immune system evolved for you to do that." After all, we humans have mostly lived in a much microbially richer (that is to say, dirtier) world than the unnaturally clean one most of us find ourselves in today.

"I think there are probably certain health conditions—particularly around the gut—where just the consumption of live microorganisms is probably an enormous benefit," Mills says. "We've evolved to be exposed to a lot of bacteria—even pre-fermentation era, I would predict, because food wasn't cleaned up to the extent it is now," he says. "And probably our immune systems have evolved to expect large numbers of microbes—and to sift through them, if you like, and make sure that they detect danger or respond to appropriate changes in diet," he says. "We've evolved to eat large numbers of live organisms. We don't do it enough anymore."

For his part, Colin Hill goes so far as to suggest that there should be dietary recommendations for daily consumption of live microbes. "I'm not against processed foods, but I think the evidence seems to be mounting. You're just losing a lot of the live bacteria in processed foods." And these are bacteria our bodies have evolved to expect, he says.

Of course, we will not be returning all of our food chain to a pre-Pasteurian model. We would like to continue to keep our milk typhoid free and our beef, sausage, and cured meats free of botulism. "Obviously we need to properly eliminate pathogenic organisms in our food," Hill says. "We can't just expose people to the plethora of organisms. We have to be clever," he says. One way to do that is, of course, to eat food that has been fermented to offer an array of nonharmful microbes.

• • • • •

Diet, scientists have discovered, can also go deeper than simply providing transient microbes or food for native microbes to feed on. It can actually change the genetic landscape of our microbiome. This is possible because bacteria have a much looser relationship with their genetic code than we do. Bacteria are, shall we say, genetically promiscuous. They can swap bits of genetic code with neighbors (without the same amount of awkwardness that might ensue in the human realm), trying new traits on for size. (Need to be able to digest gut mucus? Perhaps your buddy over there has a code for that. Struggling to get by after your host has taken a double dose of antacids? Check with that new arrival drifting by. And voilà—old microbe, new tricks.)[†††] When you combine this propensity with the rapid rates of replication bacteria have, you have a recipe for, as one scientist described it to me, "opportunistic drifting." Some proponents of live fermented foods even cite this as a reason to continually freshen up the gut's communal gene pool with new ingested microbes—and their genes.

One aspect of the Japanese diet has altered local human microbiomes through a fascinating example of this bacterial gene-swapping: seaweed. Most humans and their microbe collaborators are unable to digest one of the prevalent carbohydrates in seaweed. Over many millennia, however, Japanese populations have developed this capability—thanks to their common microbes. Their guts harbor a bacterium (*Bacteroides plebeius*) that manufactures enzymes that can disassemble this regionally common food and obtain extra nutrition for the host from it. That is all well and good, but the curious thing about this arrangement is that the genes responsible for making this useful enzyme are not at all native to this type of bacterium. Instead, the necessary splice of genetic code comes from an unrelated ocean-dwelling microbe (*Zobellia galactanivorans*), which lives (and feeds) on seaweed in the wild. So helpful was the ability to glean additional nutrition from the seaweed that the human gut bacterium borrowed this section of code for its own book of base pairs—sending it down through the generations of bacteria and people who found it useful. So this is just another reason to seek out the wild and lively microbe-covered foods. You never know what helpful genetic tricks your microbiome might pick up.

††† This capability, however, is also the reason for the rapid and worrisome rise of antibiotic-resistant strains. If one microbe harbors a genetic mutation that allows it to survive a blast of antibiotics, it can fairly easily spread that bit of genetic code to other (sometimes harmful) microbes, endowing them with these dubious superpowers.

Better Together

Once scientists discovered probiotics and then prebiotics, they (and food and supplement makers) wondered what would happen if we could harness the power of both of these types of foods. Enter synbiotics (the *syn* of *synbiotic* is meant to invoke the idea of synergism). These foods provide beneficial microbes and the foods they like to eat.

But as is true of many areas of this field, a little nuance here often gets overlooked—willingly or accidentally. Just as you can't give any vehicle—car, plane, or space shuttle—any type of fuel, not all types of prebiotics "work" for all types of beneficial microbes. So using an oligofructose with a *Bifidobacterium* strain that can consume that prebiotic would qualify as a synbiotic because it encourages the growth of that *Bifidobacterium*. Conversely, an oligofructose with a *Lactobacillus casei* strain that doesn't feed on oligofructose compounds would not. But rather than get caught up in these details, many naturally fermented foods are already synbiotics. Kimchi and sauerkraut are just two types of these extra gut-friendly foods.

As many scientists have concluded, these elements of diet are about shifting the balance, stacking the odds. Sure, probiotics are transient. And prebiotics can help boost only the microbe species that are already in your gut. But even in this limited context, there seems to be plenty to gain and not much to lose.

When it comes to our guts, we need to shift our mindset from a model of growing crops—or even planting a kitchen garden—to fostering a highly diverse, complex ecosystem that we might not yet be able to fully understand and interpret. We need to foster our own individual wild jungles.

And the answers undoubtedly lie not at the bottom of a plastic probiotic yogurt cup. Or in a jar of capsules in the refrigerator. We should be looking, instead deep in our cultures, where fermenting food and eating large amounts of fiber were common. It's doubtful that simply adjusting a few aspects of our diets will cure chronic diseases or prevent them from occurring in the first place. But by learning more about how foods nourish us and our microbes, we can begin to better understand how the act of eating affects our health, with microbes as the mediators. Those thousands of years of culinary trial and error and subtle dietary tweaking can't have been for naught. Our ancestors' cuisines were constrained by a myriad of other forces—of climate, ecology, local know-how—as well as by the nutritional requirements of our own bodies. But these ancient cuisines that emerged must have been doing *something* to also nourish a flourishing microbiome. Otherwise, we surely would have perished. Until recently we knew nothing of our microbiome's depth, its importance, or even its very existence. But as a species we have certainly developed rich and diverse ways of feeding it.

It's time we (re)learn how to feed our bugs.

The most effective—and delicious—way to do this is to dive into the live and surprising world of microbiome cuisine: the ancient, the strange, and even the cutting edge. And the best places to do that are at the source, the locales where these foods emerged, thrived, and have been perfected over generations. The places that still make many of these foods in much the same way they did before the commercially processed facsimiles made it into our grocery stories. The places where food is still wild, dirty, powerful, and enticingly varied. Places like a high mountain village in Greece, the urban restaurants of Seoul, a seven-hundred-year-old miso company in Japan, and a musty cheese cave in Switzerland. Along the way there will be microbe-enriching local cuisine to eat (and digest), traditional cooks to learn from, and scientists on the leading edge of research to ask questions of. Food preparations and an exploration of the latest science on the microbiome will help bring many of these discoveries home to your plate—and to your gut.

CHAPTER FOUR

Quintessential Culture

Dairy

In the current popular lexicon of probiotic foods, one looms larger than all the rest: yogurt.

For decades, many small brands of yogurt have alluded to their live and active cultures, whispering to the health food-seeking subculture with small-print enigmatic advertisements of obscure actors: *Lactobacillus acidophilus, Lactobacillus bulgaricus, Streptococcus thermophilus.* Incantations of some dark, unseen magic.

Since the early 2000s, as science has caught up with microbes and their links to health, big brands have seized upon this convenient snack food as a hot ticket.*

The possible health benefits of yogurt, though, far predate daytime television commercials and even those earthy 1970s health food shops.

---

* Since then, some have made financial amends for going *too* far in their promise of good health; Dannon has settled lawsuits for tens of millions of dollars over exaggerated health promises about its Activia yogurts.

In fact, its health lore begat the entire scientific study of probiotics. Yogurt origin stories and folk remedies can be found in cultures the world over, from Scandinavia to the Middle East to the Indian subcontinent. Persian legend has it that Abraham achieved his impressive longevity (one hundred and seventy-five years) and fecundity (fourteen sons) thanks in large part to his regular consumption of yogurt.[†] And in nineteenth-century Eastern Europe, a renowned scientist first took notice of this humble dairy product, known there as sour milk.

A Russian Scientist and the Quest for Longevity

The story starts with a brilliant, bearded Russian scientist named Élie Metchnikoff. Among his many discoveries, one stood out. He found that Bulgarian peasants had one thing that he—and the rest of the world—wanted: longevity. As a trained biologist, he focused a keen eye on their lifestyle and diet, eventually zeroing in on their frequent consumption of cultured milk.

Metchnikoff was born in what is now Ukraine and spent his early scientific years studying the unglamorous nematode. By his thirties, he had turned his attention to the immune system. In 1888, he arrived in Paris to work under the renowned Louis Pasteur at the latter's famous institute. There Metchnikoff made his future Nobel Prize–winning discovery of a new type of immune cell, the phagocyte. He later shifted his focus yet again, this time to aging and longevity.[‡]

† A claim that is perhaps not entirely off base. A few years ago, scientists looking into yogurt's effect on obesity made a serendipitous discovery that male rats that ate yogurt had discernibly larger testicles and were able to more rapidly impregnate female mice. A team of Harvard researchers is looking into whether yogurt might have similar effects in humans.

‡ He is even credited with coining the term gerontology.

He noticed that an unusual number of people in some areas of Eastern Europe were living beyond one hundred (roughly double the average life expectancy at birth in Europe and the United States at the time)—despite their very modest circumstances. What intrigued him most of all was their frequent consumption of what they called sour milk.

From his research, Metchnikoff was convinced that the process of aging was spurred on by bad bacteria that accumulated in the large intestine (an organ he called a "vestigial cesspool"; even he was not entirely prescient). These bacteria, his thinking went, released toxins into the system. The toxins in turn prompted the body's immune cells to attack healthy human tissue—from nerve cells in the brain (leading to senility) to color-producing cells for the hair (creating white hair). The concept fell right in line with the ancient (and surprisingly long-lived) notion of "autointoxication," the idea that waste sitting in the colon poisoned the body from the inside, leading to a host of medical issues from headaches to hysteria. This popular belief, which persisted into the twentieth century, led to all sorts of, as one researcher put it, "colonic quackery."[§]

But Metchnikoff came to believe that lactic acid bacteria, such as those found in the Bulgarian dairy products, could counter these negative effects, extending the life span. He knew that lactic acid bacteria used in fermented milk products could stop the growth of so-called putrefactive microbes through their production of pH-lowering lactic acid. One promising organism he singled out was "*Bulgarian Bacillus*"[¶]—now known as *Lactobacillus delbrueckii* subspecies *bulgaricus* (or sometimes just *L. bulgaricus* on food labels)—which he believed take up

§ And, along with it, the sale of lots of colonic irrigation products. And even some rather radical surgeries to remove the colon altogether. The testimonials were apparently very convincing, the science less so.

¶ The bacterium was first isolated in 1905 by the Bulgarian microbiologist Stamen Grigorov, the same microbiologist who would later go on to develop a vaccine treatment for tuberculosis that used penicillin-producing fungi.

residence in the gut, banishing those bacteria he saw as toxic by creating an environment too acidic for them to survive. So convinced was he of this idea that he began drinking milk cultured with this bacterium daily.** Soon he had other doctors on board, too, and they began prescribing sour-milk diets to their patients.

Metchnikoff's pioneering ideas turned out to be incredibly influential, if slightly off base. Writing in 1907, he noted that "the dependence of the intestinal microbes on the food makes it possible to adopt measures to modify the flora in our bodies and to replace the harmful microbes by useful microbes." As we would learn later, his yogurt microbes, alas, don't actually take up residence in the gut. Nevertheless, his theories about the potential benefits of particular microbes acting in the gut helped spawn the quest for probiotic bacteria and the study of the microbiome. His 1907 book *The Prolongation of Life: Optimistic Studies* spread the word about yogurt's health benefits around the globe for decades to come.†† Even today, yogurt marketing harks back to this healthful ethos.

• • • • •

For the first part of its history in the United States, however, yogurt was not the commercial success it is today. It was mostly sold quietly in communities of immigrants as a specialty food, purchased alongside other tastes of the homeland. Thanks to Metchnikoff's work, in the

** Alas, he did not join the centenarian club, but he did surpass the expected life span of the time (despite two earlier failed suicide attempts, each after the death of a wife from infectious disease); he died in 1916 at the age of seventy-one, from heart failure.

†† Metchnikoff's notions persisted through at least the mid-twentieth century. As the authors of the timeless *Joy of Cooking* note: "The longevity of certain groups of Arabs, Bulgars and other eastern peoples is often attributed to their diet of sour and fermented milks. The friendly bacteria in these milks settle in the intestines."

early twentieth century, it began to gain some popularity as a treatment in fringe medical circles.[‡‡] Still, it was not achieving much of a market foothold. Around that time, however, across the Atlantic, a young man named Daniel Carasso, the son of a yogurt maker, left Spain for France to study business. The young entrepreneur then took up bacteriology at the Pasteur Institute. Thus armed, he took the reins of the company his father had started, called Danone, after Daniel's own childhood nickname. When the family was forced out of Europe altogether by the Nazis, Daniel immigrated to the United States. One there, he and two partners bought a small Greek yogurt factory in New York City. There they changed the name to the more American Dannon.

At first their product didn't exactly fly off the mainstream shelves. Customers were primarily those of Greek, Turkish, or Middle Eastern origin, who were familiar with the tangy product. To help get the word out, the businessmen hired a New York City advertising firm. And out it got, though not in quite the way they might have hoped. According to company lore, yogurt rapidly became the butt of popular jokes.[§§] But two things changed all of that. One was *Reader's Digest*, and the other was fruit.

In 1947, the sour food was still failing to gain widespread favor among New Yorkers. So the founders happened upon the idea of adding strawberry preserves to the jars, making the yogurt sweeter and more appealing. Sales began to pick up. Then in 1950, an excerpt from the soon-to-be-published book *Look Younger, Live Longer* ran in *Reader's*

‡‡ The famed sanatorium director (and creator of cornflakes) John Harvey Kellogg became quite enamored with the substance. His fixation on the bowel as the source of most ills fit nicely with Metchnikoff's work on toxic versus protective bacteria. Kellogg subjected many of his patents to regular bowel cleanings (via extreme hydration and enemas) and then aimed to replace the microbial communities with those from yogurt, also delivered via both ends.

§§ One, as legendarily retold by one of the founding partners, goes: "There was one about the ninety-seven-year-old woman who died—but the baby lived."

Digest. In his book, author and health food fanatic Gayelord Hauser praised yogurt as "a 'must' among the wonder foods." Soon the (now-sweet) food became a common American snack. Carasso eventually moved back to Europe a successful businessperson. We can't say for sure if it was the yogurt, but Dannon's namesake lived long enough to see the company's tremendous popularity, dying only in 2009—at the age of one hundred and three.

• • • • •

Nowadays, as consumer preferences have shifted, yogurt producers have diversified from the mixed-fruit variety to an explosion of flavors in spoonable and drinkable consistencies—and then bounced back toward the plain and tangy and the firm. Today the dairy shelves of my local supermarket are filled with more than a hundred linear feet of yogurt products, which is not unusual in this hemisphere; yogurt is now a $6.7 billion industry in North America alone.

In the Thick of It

Yogurt wouldn't be yogurt without those little ingredients that are too small to see: the microbes that make up its culture. Unlike many other fermented foods, such as pickles, which ferment spontaneously, classic yogurt requires the addition of specific microbial cultures. What we think of as yogurt comes to us courtesy of two hardworking bacteria: *Streptococcus thermophilus* and the familiar *Lactobacillus bulgaricus.*

To make yogurt, one first needs to prepare a substrate in which the [illegible]s will flourish. Heating the milk first alters the protein structure [illegible]s off other milk-dwelling microbes that might compete with the yogurt makers that will come in next. The bacterial cultures are then added and given time to multiply away. The final yogurt is the result of the bacteria's metabolic processes as they eat up compounds from the milk and produce acids as by-products. Remember the beneficial acids manufactured by our gut bacteria? The same principle applies here—and in most fermented foods.

In yogurt, the bacteria take in lactose and other compounds from the milk, and from them, produce lactic acid and ethyl alcohol. These two main bacteria—*S. thermophilus* and *L. bulgaricus*—work together to make the surprisingly nuanced process smooth and speedy. *S. thermophilus*, which is the less picky of the two organisms, grows quickly first and stimulates the growth of *L. bulgaricus*. As the pH dips, the preponderance of *S. thermophilus* drops and *L. bulgaricus* becomes dominant, controlling the final phase of fermentation. Once the yogurt has arrived at its desired taste and texture, it must be cooled to stop further fermentation and souring.

If the actual creation of yogurt requires only two bacteria, what then of the brands that boast four, eight, or a dozen "live and active cultures"? Those labels that are peppered with other italicized obscure names, such as *Lactobacillus rhamnosus* and *Bifidobacterium bifidum*. These bacteria don't have much to do with the actual yogurt making process and are usually added after the fact, for marketability and health benefits. Yogurt can even be pasteurized post-culturing and then reinoculated with any microbes a company might want to add. It becomes a sort of aqueous vehicle for beneficial bugs, regardless of their role in the actual yogurt creation. By the strictest of definitions, standard yogurt—even with live cultures intact—is not necessarily probiotic. *L. bulgaricus* and *S. thermophilus* are classified as starter cultures, not as probiotics per se. And in any case, to be proven as probiotics, a microbe must be analyzed down to an individual strain; such detail that is usually not provided on the labels of "probiotic" products.

Picky definitions aside, the workhorses *L. bulgaricus* and *S. thermophilus* have been shown to donate enough of their own lactose-digesting enzymes to the consumer. This extra boost can help those sensitive to lactose digest the compound, improving digestion and possibly helping some people get more calcium in their diets by making dairy's nutrients more accessible. Even before they are ingested, these bacteria eat up lactose in the milk, reducing the overall lactose in yogurt by some 20 to 30 percent. Some of the bacteria in yogurt can also increase the amount of folic acid, niacin, riboflavin, and other beneficial compounds in the dairy product. Beyond the digestive tract, these microbes and their yogurt can really do a whole body good.

• • • • •

Originally, yogurt was not just a pristine blend of pasteurized milk and specific bacteria. Today's commercial yogurts are produced in massive, gleaming factories, where the temperature and components of the milk are carefully controlled and the precise strains of bacteria are added at just the right moments. Yogurt—and certainly Bulgarian sour milk—was not always made this way. It, along with most fermented dairy products, were created through a process known unappetizingly as backslopping.

This method involves saving a little bit of an earlier batch of yogurt to add to a new batch of warmed milk; thus a chain of souring, evolving bacterial communities is created. This practice is still how local and homemade yogurt often gets made. In Armenia and Georgia, for example, the local version of yogurt, known as matzoon, is made from cow, buffalo, goat, or sheep's milk. In many settings, it is still made by backslopping culture from a previous batch. In Mongolia, a fermented milk known as tarag is fermented by backslopping an earlier batch into fresh cow's milk, giving it a rich microbial profile.[¶¶] Even closer to home, this is how many home yogurt makers create their product. They start with a store-bought yogurt with live and active cultures (the more, the better), add a bit of it to warmed milk, allow the cultures to multiply, and then enjoy the sour homemade product—saving a little with which to start the next batch, of course.

Quest for Culture

For all of yogurt's scientific and traditional nuances, to truly understand the lively and intimate world of this, the ur-probiotic-food, I set out on a journey to the modern-day mecca of yogurt: Greece.

In Greece, yogurt suffuses the cuisine and food culture in a way that is difficult to appreciate until you arrive and start eating with the locals. Street gyros are slathered with yogurt-based tzatziki sauce. In the shadow of the Acropolis, a tony Athenian storefront serves cups of plain yogurt with your choice of toppings, where the classic honey-and-walnut order was tops. At an island pension, I'm served a traditional homemade breakfast yogurt parfait layered with fiber-filled oat bran, cinnamon, and honey. It has even been suggested that the phrase *land of milk and honey* is more accurately rendered as *land of sour milk and honey.*

Most of the yogurt eaten by Greeks today comes from the industrial giants, where the product and culture strains are standardized and carefully controlled. But especially in the countryside, plenty of families still rely on small, local producers or even make their own. It was this quest

¶¶ In one study of seventeen samples of tarag, researchers found forty-seven different bacterial genera and forty-three different fungal genera—a far cry from the handful of species in commercial yogurt.

for homegrown yogurt culture that led me to climb into the car of a tall, earnest Greek man double-parked in front of my Athens hotel early one morning.

George, the man who is waiting for me, is impeccably dressed and speaks proper, well-articulated English. A pair of local researchers who are investigating traditional Greek foods for novel microbes had arranged our meeting. George and his family are in the cheese business (which in Greece means feta), but Demetrius, a close family friend, makes yogurt. With that small connection, he points his car out into the early Athenian rush hour traffic, and we are off.

As we leave the central city and turn onto the historic Greek Marathon route, George explains that the Attic region of Greece, despite being the seat of Athens, the bustling present-day capital, is also home to a substantial portion of the country's dairy farmers. Some of their production is sold as milk, some becomes cheese, but a lot of it is used to make the country's yogurt.

As the famous road stretches out ahead, rolling hill after rolling hill of unambitious sprawl—tire shops, salons, bakeries—we chug ever closer to a place where the past meets the present: Marathon, Greece. Here, Olympics and legends aside, something quietly heroic is still taking place each day: the creation of traditional Greek yogurt.

Our destination is a stand-alone shop that could be mistaken for yet another bakery or market. Set at an angle to the main road, it does not reveal much of its full purpose. Outside, we meet Demetrius, a stocky young man with a friendly demeanor. His parents started the local yogurt operation, Daphne,*** in 1974. Demetrius's father is the owner, and his mother runs the storefront. The moment you step into the door of the simple shop, you can smell the culturing milk, a somewhat sweet,

*** In Ovid's Metamorphoses, the Greek mythological figure of Daphne is known for her metamorphosis. Apropos.

damp, but not unpleasant odor. The shop looks like an old-fashioned ice cream shop or a bakery in an outer Brooklyn neighborhood. White beadboard covers the walls and ceiling and is decorated with shepherd's crooks and pails, serving as decor and a reminder of their product's local origin. There are two simple tables with plastic tablecloths and a couple of chairs. In the glass case, where one might expect to see pastries, are rows and rows of plastic yogurt cups and terra-cotta dishes brimming with fresh yogurt—a traditional way of maturing and selling the product (here and in many other countries around the world). Next to the cash register, instead of the straws and sugar packets of a bakery, there are small plastic sampling spoons and packets of honey.

A nondescript side door from the shop serves as a portal to the cradle of yogurt production. A handful of people—including Demetrius's aging father, Geórgios—is responsible for the entire process, from warm milk delivery to cooled yogurt. The well-worn space is lined with old-fashioned mixers, milk collection vats, and racks of terra-cotta jars.

The first step in the process is collecting milk from local sheep. In Greece—and in many regions around the globe—dairy means not cow's milk, but instead that of sheep, goats, or even horses, camels, or water buffalo (a serious milking job if ever there was one)—essentially, milk from whatever large domesticable mammalian herbivore thrive in local conditions. In Greece, for example, cows don't fare well on much of the country's rugged terrain, but goats and sheep flourish. Daphne's very local milk supply comes from herds of sheep roaming the hills within about a twelve-mile radius of their factory. And that's in the off-season—it comes from even closer the rest of the year.

The next step is to prepare the milk. When we arrive, a recent delivery of milk from a local farmer is being heated in a large open vat, and the entire space becomes stiflingly hot and humid. This heating process serves two purposes common to many cultured dairy products: It transforms the protein structures, making them more amenable to creating a thicker final product, and it kills off potentially harmful bacteria, while also clearing the microbial stage for the desired introduced cultures to perform their magic.

Next the yogurt makers cool the milk slightly, letting it rest at a toasty 113 degrees for about three hours before stirring in starter from the previous day's batch—backslopping it, that ancient method of perpetuation. The yogurt takes two and a half to three more hours to mature, and then it is cooled for a day to slow the fermentation process until it is ready to eat.

Daphne's bacterial cultures are the same species now found in contemporary yogurt across the world: *L. bulgaricus* and *S. thermophilus*, although these are hardly freeze-dried additives from a lab. When I ask where they get their specific culture strains, Demetrius says simply, "From another yogurt." When I press to find out where the culture for *that* yogurt came from, he can answer only that "it goes back, back, back." So precisely which strains they are using is a mystery. "There are thousands of strains" throughout Greece, Demetrius says. And although university researchers are doing their best to collect, catalogue, and study the wide diversity of microbes—from the mountains of the mainland to the small Greek islands—the task is monumental. Whatever ancestral microbes are at work in Daphne's yogurt, they know what to do, and they do it well, working together to digest the lactose in the sheep's milk and produce lactic acid, thus lowering the pH. Demetrius usually aims for a pH of about 4.8 for their yogurt, which gives it a shelf life of about twelve days in the refrigerator.

To make their thick, strained yogurt, they place the finished yogurt in cotton bags and then press it to remove the excess water. Or they place it in terra-cotta pots. "The ceramic matures the product," Demetrius says. The clay absorbs some of the moisture, "so after a few days, it seems like strained yogurt," he explains. These pots are especially popular around Greek Orthodox Easter, which is just a week away when I visit. Families often will buy a kilogram of yogurt for the big Easter meal, where yogurt has long been a mainstay (as mother sheep, without their lamb—which is dinner—have extra milk to give). Or families will buy a small container and use it as a starter to make their own yogurt at home for the holiday. Just decades ago, a trip to the market for a starter wasn't even necessary. Yogurt was often delivered, sellers toting around a big pot and selling it by the scoop.

Traditional Greek yogurt develops a firm top, and Daphne's is no different. When that top is cracked, whey pools in the valleys, causing the jagged top to resemble a pocked planetary surface. Sheep's milk yogurt contains more solids from cream than cow's milk yogurt does. It has a distinctive flavor that brings the word *barnyard* to mind—in a good way. The family team also makes a 2 percent yogurt, which is creamier than many low-fat versions in the U.S., and unlike many large industrial brands, they don't add gelatins or other thickening agents to their yogurt, instead relying on simple homogenization. The resulting yogurt has a smoother texture and a more subtle flavor, which many Greeks now prefer to the rugged non-homogenized whole-milk version. With a little honey from the handy packets, it is indeed pretty much perfect.

• • • • •

There is still more to yogurt's creation than what first meets the eye—and tongue. More magic is to be found in the fermentation process within the metabolic process of the bacteria. The bacteria take in lactose and other compounds from the milk and produce lactic acid and ethyl alcohol. In yogurt, *Streptococcus thermophilus* kicks off the fermentation. As the pH dips near 5, *Lactobacillus delbrueckii* subspecies *bulgaricus* takes over. It's possible to make yogurt by adding these microbes separately, but together they do the job much faster.

Although these two bacteria are behind most yogurts, no two types are exactly the same. Some of that boils down to different milk bases (fresh milk versus concentrate versus powdered, cow milk versus sheep versus goat), different heat treatments, and different preparations. But a lot of the difference also comes down to the strains and their particular genetics. Different genes can alter bacteria metabolism and enzymes, affecting the finished product.

Many consumers today prize mildness of flavor instead of the sharp tang of sourness. So large commercial operations carefully control all of the variables—especially the strains of bacteria that they use. As the authors of *Manufacturing Yogurt and Fermented Milks* note, "The selection of starter strains becomes critical in obtaining mildness"—which translates to a pH somewhere around 4.2 to 4.4. Strain type is also key for the consistency of the yogurt. "To obtain a smooth texture without separation, starters containing strains that produce exopolysaccharides are necessary, but there is a fine line between the smoothness desired and stringiness or 'ropiness,'" they write. Or starter strains can be chosen for their ability to create a thickness that will hold up pieces of fruit—or for their ability to resist pH changes and oxidation that can blanch fruit. These two classic species turn out to have a world of options to offer.

By now you might have gotten a sense for the specificity and highly controlled nature of most of the commercial yogurt made today. Far from the backslopping and open-air processing of this food throughout history, today's yogurt culturing "requires specialized knowledge of microbiology, microbial physiology, process engineering, and cryobiology," note the authors of *Manufacturing Yogurt and Fermented Milks*. Despite these requirements for the creation of store-bought yogurt, making yogurt at home is far simpler. And you don't need a degree in microbiology.

• YOGURT •

To make your own yogurt, all you need is milk (whole produces the thickest results), a yogurt starter (which can be just a spoonful from a store-bought "live and active culture" yogurt), and a stable source of warmth. This is culturing at its easiest. Make as big a batch as you want. Think about how much yogurt your household will go through in a week or two. A good target quantity is a quart or two. A yogurt culture can keep on giving, so don't forget to save some for the next batch. Eat your yogurt with granola to add some prebiotic fiber to the meal.

Pour your milk into a saucepan.

Heat the milk gently to just below boiling.

Pour the hot milk into quart-size glass jars. If you wish, you can sterilize these jars in boiling water first, although many home yogurt makers simply hand-wash theirs.

Allow the milk to cool just until it is comfortable to dip your finger into. (This is a guide you can often use when gauging the survivability for beneficial bacteria: think the temperature of the human gut.)

Add the starter. If you're using another yogurt as a starter, roughly a couple of tablespoons per quart of milk will do. Give it a gentle stir.

Cap off your jars with lids.

Now it's time to incubate your yogurt. Some people use an oven with just the light on; others use a picnic cooler warmed with hot water bottles; still others swear by simply leaving the jars in a large pot of warm (again, not hot) water on the counter; and some use dedicated gadgets like yogurt makers or Instant Pots. Incubation time will vary depending on the method and temperature—just a few hours to firm up for the warmer ones, with the cooler ones taking overnight.†††

Once the milk is thickened to yogurt consistency, it's done! Place the jars in the fridge until you are ready to enjoy.

---

††† As Irma Rombauer and Marion Rombauer Becker note in the *Joy of Cooking*: "Yogurt has the added idiosyncrasy that it doesn't care to be jostled while growing, so place all your equipment where you can leave it undisturbed."

Theoretically, yogurts should be able to be perpetuated indefinitely. However, as expert fermenter Sandor Katz cautions, when looking for starter yogurts, even all-natural "live and active culture" yogurts from the grocery store are not as robust as traditional yogurt cultures. "The first decade of my yogurt making, I was just going to the store and buying Stonyfield or Brown Cow or something and starting yogurt from that," he tells me. "I never understood why, after two or three generations, I had to go back to the store and get more starter. It was like, 'How did this food even survive through the [human] generations if I can only get it to go for three [yogurt] generations before I have to go back to the supermarket to get another starter?'" He learned from microbiologists that commercial yogurts are made from only a few isolated bacterial strains. This closely monitored industrial process likely originated "because microbiologists of a hundred years ago just could not believe that a community of bacteria could be safe and desirable to work with," he explains. "There was always this idea of trying to isolate out what's really the essential components of the ferment. They isolated out of a traditional Bulgarian culture *Lactobacillus bulgaricus* and *Streptococcus thermophilus*." Thanks to that early investigation, these are the bacteria that now define yogurt for the United States, the European Union, and the UN's Codex Alimentarius for international trade. "Traditional yogurt cultures are much more complex, and they have a structure that includes defense strategies so they can survive through the generations." For his personal yogurt making, Katz notes, "I finally, through the magic of the internet, got ahold of an heirloom starter culture. I've made a hundred generations with the same starter."

Far, Wide, and Wild

Before it arrived on grocery store shelves, and even before it was being made in small batches in family shops in Greece, yogurt was being made by hand. For thousands of years, it has been crafted this way all over the globe. It is thought to have originated in the Neolithic period, either in Central Asia or the Middle East. Without refrigeration—especially in warm climates—fresh milk does not stay fresh for long, souring spontaneously from bacteria in the environment and in the milk itself. So, as they might have said in 5000 BCE, before life gives you sour milk, make yogurt.

Scholars suggest that fresh milk might have been stored and transported in animal skins or stomachs, which themselves contained bacteria to jump-start the fermentation process. Storing the souring milk in animal hides also would have allowed some of the extra moisture to drain out, further concentrating the milk and lactic acid to produce a thicker product that might be closer to what we would recognize today as yogurt. Another early practice would likely have been to store the milk in earthenware containers. As at Daphne in Greece, the clay draws out the moisture, concentrating the contents. Clay also serves the dual purpose of helping to keep the yogurt warm during culturing and then cool during storage, which prolongs the shelf life. This method is still practiced in many areas far beyond Greece, including India and Nepal.

It's hard to say what the original microbial fermenters were in the earliest days of yogurt, though surely there were many. According to its genome sequence, *L. bulgaricus*, the bacterium that so fascinated Metchnikoff, appears to have evolved to live on plants. Over time, however, as it was incorporated into generations of fermented milk products, we humans essentially domesticated it. Since then it has evolved to become exceptionally efficient at digesting lactose.

Bulgarian yogurt, or sour milk, is thought to have originated some six thousand years ago with people living in the region then known as

Thrace. Ancient Greeks consumed yogurt products known as oxygala or pyriate, eating them, as many Greeks do today, with honey. Some two thousand years ago, the Roman writer Pliny the Elder made reference to the making of a yogurt-like food by "barbarian tribes." Pliny the Elder also recommended it for treating stomach complaints—in line with what we now know about its bacterial capabilities. In the 1070s, the Turkish scholar Mahmud al-Kashgari‡‡‡ compiled a dictionary in which he recorded the first known formal description of yogurt. Less than two centuries later, Genghis Khan is said to have fed his troops on horse milk yogurt, bringing the food to people in regions he conquered.

Today yogurt and similar products are found in the traditional diets of Icelandic herders, Indian farmers, and Iranian tribesmen. According to one survey, the largest yogurt consumers in the world these days are the Swiss (who eat roughly 63.5 pounds per person each year) and the Saudi Arabians (ingesting some 48.7 pounds per person annually). But that doesn't mean the rest of us are slackers. Across the globe, people spend more than $50 billion on store-bought yogurt each year—and many others are still making it at home or getting it from neighbors or markets.

• • • • •

But to back up for a moment, when we use the term *yogurt* to describe all of these products, we are committing a gross oversimplification. The food is so diverse and rich that this would be kind of like calling all baked goods *bread*.

In India, for example, the most common yogurt-like food is dahi, which dates back at least twenty-five hundred years (some say even to the deity Krishna's time, some five thousand years ago) and is

‡‡‡ Al-Kashgari was himself rumored to have lived to the then particularly old age of ninety-seven.

mentioned in sacred Hindu scripture. In traditional Indian Ayurvedic medicine, dahi has long been thought to be beneficial for health, especially for intestinal complaints. The bacteria involved in creating dahi (outside of industrial operations) are varied, but the most common are Lactococci and Leuconostocs (also found in sauerkraut), which give it a more buttery flavor than Western yogurts.

Dahi is also the basis for lassi, the popular Indian drink in which water and various fruits, sweeteners, salt, or spices are mixed in. When blended with turmeric, it was often used as a remedy for stomach ailments.

LASSI

You can use homemade yogurt or a store-bought variety to make this simple refreshing drink. You will need about two cups of yogurt, cold water or milk, ice, sweetener, salt, and a blender. The possibilities for flavorings are as wide-ranging as your imagination: you can try mint, mango, cardamom, and/or turmeric.

Place yogurt in a blender.

Add an equal amount of water/ice/milk.

Add a pinch of salt, about a tablespoon of sweetener, such as honey, and any additional flavoring ingredients.

Blend until the mixture is frothy and serve right away.

A different type of cultured dairy product emerged in Iceland as early as 100 BCE and is featured in various classic sagas. Known as skyr, it is made much like yogurt. But it can include a rennet to increase its density, linking it to cheesemaking traditions. Studies have found *Lactobacillus helveticus*[§§§] and various yeasts in starters for skyr. After

§§§ Also used to create Swiss Emmental cheese, *L. helveticus* has been shown in some studies to lower blood pressure and ward off intestinal pathogens. It was named for the Helvetii people, who occupied what is now Switzerland during the time of the Romans.

culturing, skyr is traditionally put into cloth sacks to drain out excess moisture, creating a thickened product. In Iceland, skyr has embedded itself into the culture. For example, there are school-sanctioned skyr fights, and throwing skyr at politicians is an occasional form of protest. Even one of Iceland's legendary mischievous Santa-like elves (Yule Lads) that visits before Christmas each year, the Skyr-Gobbler, will help himself to the yogurt in your house.¶¶¶

Perhaps any country with a tradition of *thirteen* Santas is bound to be home to jovial residents. And Iceland is notable to many epidemiologists for its remarkably low rates of depression—especially surprising given that due to its latitude, in midwinter, there can be as few as five hours of daylight. Nevertheless, rates of seasonal affective disorder are lower there than they are in the United States and many other less extremely positioned countries. As Daphne Miller, physician and author of *The Jungle Effect,* notes, high rates of omega-3 consumption (through fish as well as through lambs that graze on local mosses) likely help to keep Icelandic brains happy. But perhaps the hearty portions of skyr and its microbes are also at work in the gut keeping the serotonin flowing.

To the east, in Finland, viili might masquerade as yogurt, but it is radically more complex, interesting—and slimy. It is not your average two-species snack. Several types of lactic acid bacteria have been identified in viili starter cultures, as have a handful of fungi, including *Geotrichum candidum* (which also develops on the rinds of some cheeses). Two things in particular set this food apart. First, certain strains of *Lactococcus lactis* subspecies *cremoris* and *Lactococcus lactis* subspecies *lactis* produce a distinct "ropy" slime within the viili. Second, its top is covered by a soft fungus layer of *Geotrichum candidum*, which, being aerobic, thrives on the surface layer, enjoying the air above while

¶¶¶ A relatively tame Yule Lad compared to the rafter-hiding, sausage-stealing Sausage-Swiper or the peg-legged one who is intent on harassing your sheep flock, Sheep-Cote Clod.

feeding on the lactic acid created by the bacteria below. The fungus also mellows out the flavor by reducing the overall acidity. It can give the top a slightly musty odor, but the blend of bacteria fermenting away underneath imbues the final product with a buttery flavor and transforms it into a rich treat prized by locals.

All of these cultured dairy products, in their native cultures, are not just a once-a-week snack or an occasional sample. They permeate the cuisine. Locals eat them constantly, often literally at breakfast, lunch, and dinner. Consequently, these diners provide their microbiomes with a steady stream of microbes. These specific microbes might not individually persist in the gut for long, but the continuous flow ensures that whatever benefits they might have while passing through are truly sustained. No colonization required.

With this frequent consumption, just how much bacteria are we talking about? By volume, about 1 percent of yogurt is made up of bacteria. So that would be a full 1.5 grams of bacteria in a single-serve cup of Greek-style yogurt. Compare that with a probiotic supplement off the shelf, which might have only a milligram of bacteria per pill, and you've got a pretty good—and delicious—deal with the real food.

Daring Dairy

Yogurt, even in its myriad forms, is of course not the only cultured dairy product out there. In the contemporary Western probiotic pantheon, the next most obvious option is kefir.

Kefir is a curious case. It might seem to be just a watered-down yogurt, but in its traditional form, it is actually an entirely different entity. In fact, scientists still do not fully understand exactly how it comes into being.

As with yogurt, kefir gets its sourness from lactic acid created by bacteria. Unlike most yogurts, it also relies on fermentation by yeasts.

These yeasts add an additional aromatic nuance, a small amount of alcohol, and a slight fizz (created by the yeasts' off-gassing of carbon dioxide, much as in beer and champagne—prosit!).

Kefir is made not from a culture of specified bacteria, but with kefir "grains," small, bulbous curds that some people compare to mini-cauliflowers in appearance. These grains are added to milk, which is then left to ferment overnight. Once the milk is "kefired," the grains are filtered out and added to the next batch of fresh milk.

Legend has it that the first kefir came from the Prophet Muhammad, who gave it to his followers. The kefir grains' secret was known only to the prophet, and if it was to be revealed, supposedly the grains' magic capabilities would vanish. Whether or not there is any truth to this remains to be seen—because we haven't yet cracked the kefir code.

We do know that these "grains" are composed of bacteria and yeasts—a SCOBY, symbiotic culture of yeasts and bacteria—held together by sugars, fats, and proteins. The grains are complex, tightly woven communities that can host thirty or more different species. The interior of the grains tends to be dominated by yeasts, and the surfaces are home to more bacterial action. One major compound in the mix is called kefiran, which is made by one of the resident bacteria (*Lactobacillus kefiranofaciens*). Researchers have also found our old friend *S. thermophilus,* as well as species of *Lactobacillus, Leuconostoc, Lactococcus,* and *Acetobacter,* as well as several different genera of yeasts. The balance of all of these microbes affects the flavor of the beverage, providing a flavor blend of green apples, acetone, butter, alcohol, and sourness—at least according to chemists.

These days, commercially mass-produced kefir comes not from haphazardly growing "grains" but from carefully measuring doses of bacteria and yeast that result in more standardized products. In the United States, that often means a blend of lactobacilli and lactococci plus a small amount of yeast (or no yeast at all), which makes it easier to control while also minimizing alcohol content. The label of the kefir brand I often bought, for example, listed a dozen or so bacteria, but not one yeast. So perhaps it really is little different these days than the probiotic-enriched yogurts sold on the next shelf.

Until quite recently, kefir making was *way* outside the confines of standardized industry. It was traditionally made in animal-skin bags—outside in the warmer seasons and indoors when it was cooler outside. It is rumored to have sometimes been kept near the door to a dwelling so that those passing could give it a kick when they walked by to keep it well mixed. Once fermented, the kefir was transferred to a storage container, and new milk was added to the seeded sack.

People have been studying the mysterious process of kefir's creation since at least the eighteenth century. Despite hundreds of years of scientific investigation and terrific advances in technology, however, we still do not fully understand the tight-knit universe of the kefir grain. "A kefir grain, when we look at it under the electron microscope, contains a vast array of microorganisms," says Colin Hill of University College Cork. "We can see their morphology, [but] we can't grow them. Many try and pull them apart and grow them on a medium in the lab, and they simply don't grow. We can *see* there's a vast array of organisms in there. We can't get them out," he says. The organisms' metabolisms are so intertwined—with a bacterium producing some of this, a yeast consuming it and perhaps feeding a different bacterium with some of that—that we, even with all of our fancy microscopes, growth mediums, and genetic sequencing, cannot reconstruct these closely evolved, interwoven structures.

And the mystery only deepens from there. How exactly are these tiny grains—which stay so tightly bound—influencing an entire sack, jar, or vat of milk? "The grain itself contains a very complex set of bacteria and yeast," Hill notes. "When we drop them into milk, the milk goes sour and sometimes even develops some alcohol and gets a bit fizzy," he explains. But most of the microbes from the grains do not seem to be circulating or multiplying in the milk itself. "What we find is that maybe one lactic acid bacteria and one yeast emerges and conducts the fermentation. Everyone else stays in the grain." Likely, those in the grain are benefiting from the food coming in from the milk, Hill says, but they don't show major signs of activity. Meanwhile, in the fermenting milk, the dominant microbes actually seem to produce *anti*microbial compounds, warding off any others from the grain that might think about moving into the expansive new and nutritious landscape. Therein lies the paradox. "Somehow, in the grain, they don't kill each other," Hill says. "Different environment, different things happen when organisms come into contact with each other." The Prophet Muhammad's secret—and the grains' magic—still eludes us.

• • • • •

Beyond kefir and yogurt, another fascinating, traditional fermented dairy product is kumiss. Although it shares some of its distinctive characteristics—effervescence, alcohol—with kefir, kumiss is a horse of a different color. Like all of our subjects here, it is created by souring milk with microbes, but the similarities pretty much start and end there. This Central Asian beverage arrives fully liquid, without a hint of curd or thickness, and its hue is an unexpected shade of gray. These features stem from its base: horse's milk.

The protein structure of mare's milk is different from that of other common dairy-providing livestock. This unique composition allows it to resist curdling or solidifying even under acidic fermentation, producing a nourishing beverage that has been prized for thousands of years.

Kumiss might have its origins somewhere in what is now Kazakhstan, where evidence of early horse domestication has been found. There are reports by the 450s BCE of Scythian tribes from the Eurasian steppes drinking the beverage as they traveled through the region, spreading it along their way to other people on the steppes and eventually to Mongolia and China (a similar drink in Mongolia is known as airag). Thirteenth-century explorer Marco Polo mentioned it as a pleasing beverage. The finished product has a simultaneously acrid and alcoholic bite. Some have called it "milk wine" and even "milk champagne" for its flavor and fermentation.

A traditional kumiss-making process calls for placing the horse milk into a sack made from smoked horsehide and leaving it there to ferment. The sack was reused for generations of the beverage, essentially backslopping new batches of fresh milk with the microbial starters needed for fermentation. If the milk failed to show signs of souring, the makers might also add fresh horsehide or a raw horse tendon or two to jump-start the fermentation process with new microbes.

Kumiss is still popular in the Central Asian region; to meet the demand for it, its production is now largely industrialized. Of course, health inspectors wouldn't likely look kindly on raw horse tendons in a product soon to be bottled and sold to the public. Gone, too, are the horsehides and the backslopping. And cow's milk is used much more often than horse's.

Kumiss has long been believed to have health benefits, including the ability to treat gastrointestinal issues, allergies, hypertension, and heart disease. In Russia, its purported healing powers extended even to tuberculosis. In the nineteenth and early twentieth centuries, the drink inspired a string of "kumiss cure" resorts. Far from fringe centers, these establishments attracted educated luminaries of the day, including writers Anton Chekhov and Leo Tolstoy. The well-heeled could check into these southern resorts, soak in the warm, fresh air, and drink bottle upon bottle of kumiss in an effort to cure what ailed them.****

Wholly Microbial

There is an entire category of dairy product that also depends on microbial action: cheese. And some types might even have some probiotic properties.

Now, before you cut into that Colby, it's important to understand that not all cheeses are good candidates for formal probiotic status. Despite the use of live microbes to culture cheeses, most of these microbes die or are killed off before the cheese reaches market shelves. But it seems that at least some of Europe's ancient cheese makers were unwittingly culturing microbes for our guts as well as for storage in their large, well-aged wheels. These early bacterial alchemists were working their magic with invisible powers to transform milk into something creamy, delicious, and, most important for them, something with a long shelf life.

The origin story of cheese is virtually indistinguishable from that of yogurt and other cultured milk products. As the story goes, thousands of years ago, herders stored milk in containers made from goat stomachs. These stomachs contained bacteria as well as enzymes similar to the thickening rennet used to curdle cheese today. When the sack of milk was carried on a long horseback journey under the hot sun, churning along the way, the riders reached their destination with a delicious discovery: cheese—or at least an edible curded product of some variety. Such a solid-state food also had the advantage of being easier to transport and store.

**** A two-week stay at one of these resorts failed to cure Chekhov's chronic tuberculosis; it did, however, send him off, according to many accounts, twelve pounds heftier.

Today, of course, the cheesemaking process looks a little different, even at its most artisanal. Nevertheless, many of these cheeses are made with some very interesting microbes. So the notion that some of these foods, such as *Emmental* and *Gruyère*, might indeed offer probiotic benefits required closer investigation. I volunteered for the difficult assignment.

• • • • •

As you drive out of Bern, Switzerland, the city's sturdy historic stone buildings soon give way to foothills and forests. Smooth country roads bring you into a land of rolling green hills bespeckled with small farms and steep-roofed wooden houses. And cows—lots of cows. Rounding a bend, the breathtaking, snowcapped Alps come suddenly into view as you descend into a small village. It is the landscape of fairy tales and postcard panoramas. It is the landscape of Emmental. In fact, it is the valley of Emmental. And its namesake cheese has been made here for hundreds and hundreds of years.

But these days, that is not quite where Emmental starts. To keep Emmental as Emmental, its microbial makeup—its culture—has been literally frozen in time. Its standard culture sits in a central government laboratory freezer. Rather than call any cheese produced in this area Emmental, the Swiss government defines precisely the milk, process, and culture to be used, ensuring that this "protected designation of origin" (PDO) product remains just that.[††††] Today it is one of about a hundred recognized types of Swiss-made cheeses whose official microbial cultures are kept in Bern, at the institute where Swiss microbiologist

†††† And in a new level of cultural protection, our guide, Ueli von Ah, a Swiss scientist, is pioneering ways to inject microbial markers into these PDO cheeses so that they can be certified by these biological signatures as being the real deal.

Ueli von Ah works. Von Ah, an enthusiastic and welcoming scientist-host, is the biotechnology group leader at Agroscope, a Swiss agricultural research group. The center, located in an approachable government complex in Bern, houses the labs, reactors, and freezers responsible for the bacteria in some of the most famous Swiss cheeses—as well as other food microbes totaling some 15,000 strains[‡‡‡‡] isolated from various Swiss products, from Tête de Moine ("monk's head") cheese to Landjäger cured sausage. We walk past the lab where many of the starter cultures are produced and past the noisy freezers that keep the cultures until they are ready to ship by mail to cheese makers.

Despite the careful keeping of cultures, however, even the most quintessential Swiss cheese is a force that cannot truly be controlled. So what I said about the microbes being frozen in time was more a technical laboratory practice. Because once these freeze-dried cultures are released to cheese makers, the product is free to make a life of its own. "They are raw milk cheeses—they all have their own bacterial flora," von Ah notes. That initial richness "makes it difficult because it's quite a complex ecosystem." Once the cultures are in the cheese maker's hands, differences rapidly emerge. The cheese maker must first wake the cultures back up and grow them during the propagation stage. To do that, the makers use milk. But even what kind of milk that is—whether raw, fresh, or concentrated—starts to determine what direction the cultures will take.

One of the cheese makers that receives these carefully guarded cultures is Daniel Stalder, who crafts Emmental in a small factory with an attached store, run with his wife, Brigitte. The shop sells not only their

‡‡‡‡ That might sound like a lot, but the largest holder and distributor of strains is not an industrious government agency somewhere, but the commercial company Chr. Hansen in Denmark. They can sell you anything from protein-building strains for cottage cheese to microbial blends for soil to increase crop productivity.

own cheese and dairy products but also sundry foods—gherkins, wine, canned veggies, and handmade yogurt. Local eggs are displayed in a large basket from which customers can select their own.

Next door, the diminutive factory room is all sparkling white tile and gleaming metal, the clear morning's spring sunlight streaming in through tall windows and doors. It is like being transported into an idealized, cheerful nineteenth-century-era factory (with twenty-first-century European labor and safety laws). A handful of young men buzz around the small space in constant motion. They wear white caps, white T-shirts branded *Emmental* in red lettering, white pants, white aprons, and even white galoshes. The only smell in the immaculate room is that of heating milk.

The milk, still warm from local cows, arrives directly after the morning and evening milkings. And then the craft begins. Cheesemaking is as much a science as it is an art. Before starting the process, the maker must analyze each delivery of milk, assessing its characteristics—fats, proteins, etc.—and what it is to become. Milk attributes, for example, change with the seasons and the cows' diet, a shift cheese makers must take into account. Stalder employs a small nook off the main room for some basic scientific measurements before he takes the milk delivery into production. Once accepted, the milk gets strained through a mesh filter to remove any barnyard debris from the creamy, opaque liquid. Its next destination is a giant copper tub that takes up much of the room's floor space.

Meanwhile, in a small walk-in refrigerator, Stalder keeps his cultures, which he propagates them the old-fashioned way: using the same milk he will use to craft his cheese. He adds small amounts of culture to one-liter milk jars and lets them multiply before introducing them into a cheese batch.

By midmorning, a vast copper tub is filled with fresh milk from that morning's delivery and is being heated. When the milk reaches the right temperature, Stalder adds the raw-milk-propagated culture, mixing it into the tub with large paddles. It is then time for the liquid to set. During this process, the milk begins to coalesce. Before it becomes too thick, the forming curd is broken with paddles. Then at just the right moment, signaled by Stalder, a lever is turned, and the massive copper tank begins to drain, sending the mixture of curds and whey flowing across the room through pipes and into small fountains positioned above seven waiting cheese molds. The mixture steams as it spouts into the molds, chunks of curd visible in the spray of whey. The liquid whey drains out of the molds and into a trough below (to be collected and fed to local pigs). It seems like an impossible amount of milk. But, Stalder notes, with von Ah interpreting the Swiss German, each wheel of cheese will weigh in excess of 200 pounds, and it takes some 1,000 liters (more than 250 gallons) of milk to create each round.[§§§§] Stalder samples curds from the side of the molds to check their texture. During all of this, the cultures are quietly eating and multiplying. And later, in the cheese cave, they really come into their own.

As soon as the molds are filled with curds, the pressing begins. Once performed with a grooved wooden press, this process is now accomplished with mechanized pneumatic presses that apply heavy pressure in just the right amount—and can even flip the molds to maintain uniformity within. The wheels of cheese will remain in the presses until very early the next morning. Overnight, bacteria help to acidify the proto-cheese, converting lactose into lactic acid, which will feed other microbes, such as the propionibacteria. The next day, the wheels are transferred to a two-day salt bath. This adds salinity—for flavor and for preservation—while simultaneously drawing out moisture from the wheel, firming it up.

---

§§§§ That's a morning's milking from a herd of about eighty Swiss cows to make each wheel.

After the salt bath, the wheels are moved in vertical racks of seven (that's a 1,400-pound tower of cheese) into a dim aging room, or cave. Here a few fans help circulate the fragrant air, which stays steady at about 70 degrees Fahrenheit. Stalder walks briskly around the room, surveying his work with a small pointed rubber hammer (much like the one doctors use to test reflexes). He taps one wheel, and it gives off a dense *clump*. "You knock on it, and you can hear that it is still very young," he says. Older wheels have a harder, sharper sound. "A cheese maker hears how the cheese quality is," von Ah explains. Stalder also measures the wheel height, which, to be a full-fledged, certified PDO Emmental, is required to be seven and a half inches.

Once the cheese has progressed enough in the first room, the staff moves it to a second room, where the temperature is much cooler, just above 50 degrees Fahrenheit. In here, there is a much more intense rich, damp, earthy umami smell. Stalder pulls out a large wooden paddle to shift wheels around and then treats us to a core sample from a wheel made two months earlier. Although it is not finished aging, it is already creamy and subtle, with a faint suggestion, a sour hint, of the product's rich future. He seals the hole back up and briefly surveys his kingdom of microbial action. As they age, the wheel rinds develop a duller color, and the older wheels here are getting small white speckles of fungal mold on their rinds. These are all signs that the wheels are en route to the full, nutty, creamy final product. After doing their time here, the wheels will be sold to a distributor for final aging and for their journey to market, where they will be more than worth their weight in euros (which, by the way, comes out to more than $5,000 retail per wheel).

By now the next milk deliveries are beginning to arrive, but they will have to wait for tomorrow. This small shop runs only one production cycle a day, unlike larger factories, which can keep the wheels turning, so to speak, all day.

In the building's cool basement, the Stalders age extra-potent

Emmental for the store. In this small cheese cave, which smells heavily of ammonia—a by-product of the aging cheese (or more accurately, of the microbes helping to mature the cheese)—the bottom shelf holds a three-year-old wheel of Emmental, still aging on wooden planks. Its surface is a textured yellow-brown, not unlike a round loaf of hearty country bread. Cheese here is wiped with salt water once a week and flipped. A regular devotional over the years, for a revered piece of microbial craftsmanship.

• • • • •

Although the cheesemaking process is ancient, we still don't have a very good understanding of what happens, microbially speaking, throughout the process. *Lactococcus lactis*, a popular bacterium for making hard cheeses, likely originated from the plant world. But after hundreds of years of use in dairy, like yogurt's *L. bulgaricus*, it has evolved to thrive in milk. It has adapted its genes to be an efficient consumer of lactose thanks to our persistent human pressure. Also, as in yogurt, in cheese, many of the bacteria work together. For instance, in Emmental, *Lactobacillus delbrueckii* subspecies *lactis* is adept at breaking down proteins in the cheese. This in turn provides the essential *Propionibacterium freudenreichii*[¶¶¶¶] with the liberated amino acids it needs to grow. From this food, the propionibacteria generate acetate and propionate (remember these from the beneficial fatty acids our own gut bacteria generate), lending to the cheese's nutty flavor. This bacteria species also creates the carbon dioxide that makes the signature "eyes" in the cheese.

This two-microbe system seems chemically sound and tidy enough.

¶¶¶¶ Strains of which have been studied for potential probiotic effects on the immune system and benefits for those with IBD.

But the microbial world of cheese is so diverse, so full of surprising bacterial strains (including two new ones recently discovered in a French Gruyère) and amazing fungal molds that we are still really just beginning to slice into the real action.

Getting to the bottom of these microbial communities is not easy, but it could hold some tantalizing answers about what really creates these cheeses. It could even lead to new discoveries about microbial evolution itself. Benjamin Wolfe, who studies the ecology and evolution of microbial communities at Tufts University, has turned his love of cheese into a scientific career. "It boggled my mind how we could have so many types of cheeses, given that they all start from the same ingredient—milk," he says. "The diversity and activities of the microbes, guided by the hands of talented cheese makers, are the sources of all this delicious diversity."

That diversity means that there are undoubtedly many complex microbial dynamics at work. For, example, Wolfe says, "When you bite into a wheel of Camembert, is that community of microbes a place of war or peace? Are microbes interfering with one another by producing antimicrobials? Are they helping each other by producing compounds that stimulate each other's growth? What happens when microbes evolve with neighbors? Do the neighbors provide new opportunities for evolution? Or maybe neighbors suppress rates of evolution. Most prior studies have looked at just one microbe living alone." And the miracle of cheese is thanks to all of these microbes living together.

Cheese is one of the many traditional fermented foods and environments that involve communities of bacteria and fungi changing together.***** So "we can learn quite a bit about how these two very

***** This nuanced study of cheese also reminds us that "you have to get away from the thinking of 'good bugs' or 'bad bugs,'" as von Ah says. For example, some starter cultures in southern Europe use microbes such as enterococci that are avoided in northern Europe, he notes.

different types of microbes interact within microbiomes," Wolfe says. "Many of the types of bacteria and fungi that grow in these foods have close relatives that live in soil and in the human microbiome. So we can translate the processes and mechanisms we learn in foods to more complex microbiomes." Wolfe and his colleagues continue to study the intricacies of these dynamics. But traditional cheesemaking, with all of its microbial mysteries, continues in its many forms.

• • • • •

Just an hour southwest of Emmental is the small, ancient village of Gruyères, which is home to, yes, Gruyère cheese. In the cobblestoned village atop the hill, at least one cheese maker shuns even the modest mechanization used by Daniel Stalder, instead preparing his cheese wheels almost entirely by hand. In a cozy storefront, he scoops out curds with a small-mesh net that he loops expertly between his bare fingers. This net, held as if to catch fish, opens to collect curds in the warm butter-colored liquid. He then lifts the contents up and into waiting molds before slowly cranking down his old wooden press at just the right pace. In the surrounding hills, many other locals still make their cheese in much the same way.

A lot of the town's namesake cheese finds its way around the globe. A decade ago, more than 2 million pounds of Gruyère were sold through grocery stores around the globe each year, and today it is surely a good deal more. On tables around this small town, you are guaranteed to encounter plenty of the local cheese. It is in the fondue, the raclette, and the soup; it is even available for slicing at breakfast. And if you're staying with a local farmer, the cheese will likely be of their own production.

Jacques and Eliane Murith run a small farm, inn, and chalet in

Gruyères, where they keep a few dozen cows. In the winter, the couple lives in a farmhouse just below the village. Their youngest cows graze in the green pasture just out back, large brass bells gently clanking near the back door to the house in the cool April mornings.

Each spring, Jacques Murith hikes with the cows—and a sanctioned Gruyère starter culture—up to the family's rustic chalet in the hills. There he will stay all summer with one of his sons, letting the cows graze on the fresh mountainside roughage, milking them twice a day, and turning their milk into the family's very own Gruyère cheese.

Jacques Murith learned the trade nearly fifty years ago from an old cheese maker, and he has been making cheese the same way each summer ever since. He starts each season with a mail-ordered starter culture. But from there, his cultures take on a beautiful life of their own. He adds the standardized starter to some of his cow's fresh milk, creating his own culture from which to start his cheese for the season. Then each day he uses cultured milk from the previous day's batch to culture the next one, perpetuating his own lineage of microbes through the timeless practice of backslopping. Microbes come not only from the cheese cultures but also from the cheese maker's bare hands and from the environment. Some researchers have even found that the microbiota of old wooden cheesemaking equipment is full of helpful bacteria that crowd out would-be pathogens. It's a safety measure as old as time.

"Food fermentations have always been a custodial process," write David Mills, the UC Davis food microbiologist, and his colleague Nicholas Bokulich in a paper on the topic. These food makers have been "intentionally or unintentionally managing microbial communities in the fermentation and the environment through direct and indirect interventions, including environment conditioning, moisture control, and cleaning procedures," they note. "This has historically been an 'artisanal' process, developing management practices through empirical trials, and these traditional methods emerged long before knowledge of the microbial denizens of these fermentations." However Jacques Murith has found to cultivate just the right balance of microbes in crafting his traditional Gruyère, it is delicious.

The process is also quite hard work. And the hardy Swiss cheese makers have developed their own humble cuisine for these summer seasons in the mountains. One of the specialty dishes that has grown out of this laborious and simple existence is soupe du chalet. Its essential ingredients are potatoes, onion, and of course, fresh milk and Gruyère cheese. "The idea is you can cook with what you have in the place—in the chalet, on the mountains," Eliane Murith explains to me in French. "Often it was just men there,[†††††] and they just stocked potatoes, milk, onions, and cheese." Cultivated or gathered greens can make an appearance, as can macaroni noodles, but to find so much as a carrot in the soup would be unusual, she notes.

A local restaurant in the historic town center, Auberge de la Halle, serves its own take on this traditional dish. The restaurant's back room looks out onto the village's ancient fortifications and the yet more ancient Alps beyond it. Inside, the soup is served in a giant wooden bowl with free refills ("as much as you want," reads the English menu, ambitiously). Cream, spinach, potato, small penne pasta, and leeks come together in a hearty but not too rich meal. A healthy side of croutons and grated Gruyère cheese are served alongside for topping.

††††† Eliane Murith, a French transplant to the village, in fact turned heads in her early years there when she decided to spend time up in the mountains with her husband.

• SOUPE DU CHALET •

When it comes to cheeses, one might not exactly need guidance as to how to consume them (sliced on a cheese board seems pretty straightforward). But there are some enchantingly simple dishes that integrate these aged fromages.

This simple entrée is flexible and can be adapted for different quantities and ingredients. You will need several potatoes, a couple of onions, a quart or so of whole milk, several ounces of Gruyère, a pat of butter, and optionally, a couple of cups of spinach or foraged greens, macaroni, and salt and pepper to taste.

Sauté sliced onions with butter in a large stockpot or Dutch oven.

Add sliced potatoes and cook until partially tender.

Add the milk with a splash of water and bring to a low boil, stirring to avoid scalding.

Cover and allow the mixture to simmer for about half an hour.

Add the macaroni and greens (if using) and cook until tender. Remove from the heat.

Add the grated cheese and salt and pepper to taste. Stir and serve right away—preferably with a chunk of whole-grain country bread.

Even a simple farm breakfast in Gruyères includes a bit of the cheese to round out the meal. Rustic freshly baked whole-grain bread is served with homemade jams, homemade butter from a decorative mold, and of course, the family's Gruyère cheese, presented in an impressive gigantic wedge. Coffee, tea, and hot cocoa are at the ready to wash it all down.

Although they might not match the latest diet-book ideals of healthful eating, all of these meals offer up a delicious combination of potentially probiotic cheese, along with a healthy serving of prebiotics in the forms of whole grains, vegetables, and cocoa.

• • • • •

Many of these cheese microbes have yet to be put through the formal probiotic paces by scientists. We know that some do survive the gastrointestinal trek. And others have shown potential benefits in small trials. But if we go back to the strict definition of *probiotic,* Colin Hill says, "that's where the line gets fuzzy. You know cheese contains a lot of bacteria. There are often lactobacilli as well as other organisms." But, he says, "they seem to be generally a different type of lactobacilli—not the type of lactobacilli you find in the gut. They've been selected for industrial and technological properties—of getting a good flavor and so on—and so they may have moved far enough away to not have sufficient probiotic effects." Still, he says, "they're somewhere on that spectrum of 'it's good to eat lots of live bugs.'" It all boils down, he says, to the question of "are the live microorganisms in cheese better for you than a pasteurized cheese that contains no microbes at all? I would think almost certainly, yes." If their rich, complex flavors weren't reason enough, that seems sufficient rationale to seek out these aged, microbe-rich cheeses.

Functional Foods

Not all cultured dairy products have their basis in centuries-old caves or animal-hide sacks. In Japan, an interest in fermented food, in health through food, and in scientific innovation led to the creation of perhaps the earliest commercial probiotic product: Yakult.

Yakult is made with powdered milk, a proprietary strain of bacterium, a dash of flavoring, and a hearty serving of sugar. This sweet blend is packaged in single-serve two-ounce plastic bottles and is now sold in more than thirty countries. Each day, people all over the world drink some 28 million of these small bottles. This manufactured probiotic drink is a far cry from a traditional fermented dish. But it has its roots in history.

Yakult began in the early twentieth century, with a Japanese microbiologist named Minoru Shirota. Shirota was taken by the work of nineteenth-century scientist Élie Metchnikoff and his quest for health and longevity through beneficial bacteria. So Shirota decided to pick up where Metchnikoff left off. As a researcher at Kyoto University, Shirota went searching for other bacteria (beyond yogurt's *L. bulgaricus*) that might help improve human health. And he found one in *Lactobacillus casei*.

Strains of *L. casei* are found in the human intestine and mouth—as well as in various cheeses, some naturally fermented olives, and other foods. Some strains are considered probiotic for their ability to prevent antibiotic-associated diarrhea and to help shorten other bouts of diarrhea. The particular strain he settled on in 1930 is now known as *Lactobacillus casei* Shirota. Starting there, he built a beverage company that centered less around time and tradition and more around industry and science. The first Yakult product hit the market in Japan in 1935.[‡‡‡‡‡]

• • • • •

Today, Yakult's main plant is located in the shadow of Mount Fuji, about two hours outside of Tokyo. When I arrive, the mountain is obscured by low clouds, a steady rain falling on the perfectly manicured campus. A group of people rush out to meet me and my two accompanying company minders with umbrellas, and together we all pass

‡‡‡‡‡ Coincidentally, around the same time that the early antibiotic Prontosil was first shown to be effective. From the 1930s on, of course, antibiotic development would rapidly outpace probiotic development.

through sliding doors and into a lobby that feels at once dated and futuristic, a humbled Tomorrowland of the functional food industry.

The plant turns out some 1.4 million consumable products per day. My private tour guide is a tall, upbeat young man, with a shining smile and well-combed hair. He is dressed for the job in a white thick cotton company jumpsuit with white shoes. The getup gives the impression of an agile fastidiousness.

The production floor here is a far cry from a musty Swiss cheese cave or even a family-run Greek yogurt operation. Here, peering down through upper-story windows in the tour-friendly hallway and into the production room, I can see more than a hundred large, gleaming metal tanks, all connected by a matrix of pipes and multilevel catwalks. Each vast tank is a towering 26 feet tall and holds some 8,450 gallons. There is certainly no backslopping going on here. In fact, between each batch, a built-in tank dishwasher sterilizes the inside—and any hard-to-reach spots are cleaned meticulously by hand by workers wearing hard hats and safety harnesses. Lest you think this an opportunity for a little bit of unknown bacterial introduction, workers, all dressed in the matching white company jumpsuits, go through an air shower to enter the production room—and a second one before they enter any tanks. Everyone receives six sets of the white suits, and workers are required to change immediately if anything gets dirty. The whole operation is a very squeaky-clean, probiotic Willy Wonka's factory.

Rather than pure imagination, however, the process here starts with powdered milk. This powder is dissolved in hot water before being sterilized. After cooling, the mixture is pumped into the pristine cultivation tanks, where it receives its precious *L. casei* Shirota seed cultures. The temperature of the tank is then raised to match human body temperature, and after a highly guarded secret "certain amount of time," the beverage is cooled, homogenized, and mixed. From there, concentrated

Yakult is shipped in shiny tanker trucks to subsidiary bottling plants, not unlike the classic Coca-Cola distribution model.

Other Yakult plants scattered around the globe (including one in the United States, south of Los Angeles) manufacture beverages for their regions. But each facility, no matter how far afield, uses bacteria shipped from the "mother strain" back in Japan.

For the stuff that is destined to be bottled right away for local markets here in Japan, the process continues under the same tight control. In the packaging room, positive-pressure airflow ensures no dust or bacteria from hallways make their way into the room. Other Yakult products beyond the classic drink feature strains that are highly anaerobic. This means that even in bottling, the liquid must be handled very carefully in order to avoid oxygen. So the company developed one machine to fill cartons without allowing air to touch the liquid—and a second machine to check afterward to make sure no bubbles sneaked in. As the tour guide points out, when you shake a container of these products, you don't hear anything—because there is no air in which the liquid can slosh around.

After nearly thirty years on the market in Japan, in 1963, Yakult introduced a new way of marketing its product: Yakult Ladies. Throughout Asia and Latin America, these women are still delivering the company's products to homes and offices. Think milkmen, but female—and for probiotics. The positions started as a way to explain the benefits of the product. But now the marketing tactic has become an institution and a key mode of distribution. There are more than 38,000 Yakult Ladies in Japan alone—and an additional 42,000 elsewhere. In addition to being a hallmark of the company, these women have also served important cultural functions. For example, after the Fukushima nuclear disaster, a scholar at Hijiyama University enlisted the help of local Yakult Ladies to gather stories, essays, and poems from affected residents as a way to share personal experiences that would have been difficult for a stranger to elicit.

• • • • •

So what is the action inside of Yakult? Studies have found that the Shirota strain, prepared in an experimental drink and taken daily during the winter, helped to prevent upper respiratory infection. Researchers have also noted that it helps boost immune function and anti-inflammatory levels in older adults.

Looking into this one commonly consumed strain has also shed light on just how complex our gut environments are—and how little we truly know about what is going on in there. In another study, participants drank a beverage with 100 million live organisms of *L. casei* Shirota for two weeks (a standard small single serving of Yakult contains roughly 6.5 billion of this bacterium, according to the company). Although the strain did not persist much beyond the consumption period, it did alter other resident gut microbes while it was there, the researchers found, changing the amounts of twenty-five other bacterial strains in the gut.[§§§§§] Additionally, *L. casei* Shirota has been shown to release a compound that hampers the ability of a pathogenic *Salmonella* strain to get around.

• • • • •

Today, Yakult isn't just manufacturing the sweetened contents of their foil-topped bottles. It has expanded into a wide variety of products and front-line research. In the intervening eighty years since Yakult was first developed, the company has expanded to numerous other functional food products, developed in the lab by microbiologists.

The company's dedicated research institute opened its doors outside of Tokyo in 1967. The 1970s saw the introduction of products containing

§§§§§ Interestingly, many of the changes, however, reduced the quantity of beneficial short-chain fatty acids that were being produced.

strains of bifidobacteria (commonly found in the guts of infants)—as well as pharmaceuticals and even lactic-acid-based cosmetics. They are now looking into prebiotic compounds to also feed beneficial bacteria. In fact, some of the early additive galactooligosaccharides (GOS) were developed by Yakult, and the company has now patented specifically developed GOS (such as Oligomate) shown to boost bifidobacteria and other beneficial lactic acid bacteria—and conveniently also double as a sweetener.

And the company has its eyes on the next frontier: space. In 2012, it launched its Space Discovery Project, aimed to study the effects of probiotics on astronauts in the International Space Station. "In outer space, astronauts are physically affected by a number of factors, such as stress arising from living in the non-terrestrial microgravity environment and confined space like a spaceship, as well as cosmic radiation," the company's literature remarks. "These factors are likely to cause changes in their intestinal environment, which may result in intestinal microflora imbalances and a lowering of immunity . . . The study group aims to verify, in outer space, the effects of taking lactic acid bacteria, which have been proved in the earth's environment." The company is also scouring the earth for new probiotic strains, including a look back into traditionally made Japanese pickles. Dairy's future as a vehicle for beneficial microbes continues to evolve.

Starting Out Gut Strong

Dairy is not only a vehicle for potentially helpful foreign microbes. It can also be a key source of food for our own microbes—starting at our very first gulps.

These are galactooligosaccharides (GOS). This category of prebiotics is found in very small amounts in most forms of milk—whether from cow, goat, sheep, or even camel or horse. They are found in the largest quantities, however, in human breast milk.

In fact, they are so prevalent that the reason for their presence has long perplexed scientists. They could clearly understand the benefit of breast milk's essential proteins, fats, and digestible carbohydrates, which are useful for immediate energy—especially for an active, growing baby. But a perplexing 20 percent of the milk's carbohydrates were not digestible by the human infant. Why would such a finely tuned, energy-intensive product such as breast milk be so chockablock with filler?

It turns out to be for the microbes, of course.

These GOS compounds encourage the growth of beneficial microbes, including bifidobacteria and lactobacilli, in the infant gut. As these good bugs multiply, thanks to the steady stream of prebiotic fuel, they leave less room for harmful invaders and can even help make antimicrobial compounds to keep the baby healthy.

Manufactured infant formula has not been able to keep up. And that's no wonder—human breast milk is not a one-size-fits-all substance. It varies from mother to mother and even in the same mother over the course of a baby's development. Among these differences are shifts in the microbe-feeding carbohydrates. Research has found more than 150 different subtypes of prebiotic compounds in breast milk so far. And these are not the same we find in other foods. They are specifically tuned, present to work with different microbes and have different effects. In addition to feeding the good microbes, they also help keep bad ones (such those that cause as strep B) out and can affect the function of the gut lining and how bacteria attach to it—or don't. Additionally, they increase the acidity of the gut, keeping invaders out and improving the absorption of crucial minerals, including calcium and magnesium, key for developing bodies.

So companies are hard at work on this prebiotic challenge. Enea Rezzonico, head of human microbiology at the Nestlé Research Center in Lausanne, Switzerland, notes that although they support breastfeeding as the best nutrition for infants, in some cases it is not possible. And for these babies, a formula that can do a better job feeding the microbes should also produce better health outcomes for the child. He acknowledges that a microbiome analysis alone can often reveal if a baby has been on formula. But they are working to erase those differences. Other companies, such as Japan's Yakult, are making inroads as well, studying compounds' abilities to stimulate the growth of some of these important bacteria, such as bifidobacteria.

Breast milk is also a medium for introducing important microbes. It is the first probiotic, if you will.¶¶¶¶¶ Breast milk introduces a very curated collection of live bacteria. Namely, *Bifidobacterium longum* subspecies *infantis*, which generally makes up the bulk of the microbes in a baby's gut early on. This microbe has been shown to be especially adept at fending off infection—a crucial service for the baby, whose immune system is still developing.

• • • • •

Mammals have been inoculating their offspring with milk-borne microbes for 100 million years or more. Beyond this initial inoculation, we humans slowly developed more methods of food preparation and preservation that continued the microbe flood into adulthood—yogurt, kefir, cheese, and plenty of others. We might never know exactly how these delicious and beneficial dairy ferments came into existence. As one UN Food and Agriculture Organization report notes poetically, "the development of fermentation technologies is lost in the mists of history."

¶¶¶¶¶ In fact, more than a probiotic. Its microbes are of human origin and are intended to actually stick around and populate the gut.

Long before these foods that required the development of careful culture curation and reliable production, there was an easier route of fermentation. It was one that can arise even without human intervention, but one that can, with some steering, can yield exquisite and surprising results.

CHAPTER FIVE

Consider the Pickle

Produce

The mainstream American pickle, often found soggying your deli sandwich or suspended in a jar of preternaturally green liquid, is a far cry from the ancient microbe-created pickle. Historically, microbes drove the pickling process, adding subtle flavors and possible health benefits along the way. Now many "pickled" garnishes are the results of a vinegar-based pickling process that is meant to destroy microbes, not cultivate them.

But not all microbe-driven pickling has been lost. There is still a rich and varied universe of microbe-fermented vegetables and fruits tucked away in home kitchens and markets waiting to be rediscovered.

Whether pickling is done with vinegar or with microbes, its aim is generally to preserve produce for consumption after the harvest. The end result is a new food with a distinctive taste—anywhere from slightly sour to intensely pucker-inducing. In the case of non-vinegar pickled ferments, they also add live microbes to a meal. And the purported health benefits from these pickles are nearly a constant: A link between pickled foods and wellness has been established in cultures across the globe. Japanese pickled foods are often said to aid in digestion and ward off illness. In the Himalayas, a popular fermented radish dish known as sinki* is thought to cure diarrhea and stomach discomfort. Fermented taro root, turned into poi, was long used in Hawaii to treat gastrointestinal issues as well as infant allergies. Although we are still learning the microbial dynamics at work in these processes—and their potential impact on our own systems—there might just be some scientific backing for these traditional preparations.

These varied dishes and potential benefits all stem from a surprisingly singular and simple process.

Controlling Spoilage

Produce is naturally crawling with bacteria and fungi from the soil, from the air, and from their own microbial communities. Once a plant is picked, its natural defenses are disabled and the microbes can take over, resulting in rot, mold, and all other kinds of processes that make fruits and vegetables less appealing for consumption.

Before the nineteenth century, there were limited ways to preserve this hard-won harvest. Fruits and vegetables could be dried, to greater and lesser success. Or they could be stored in a vinegar brine (a strong acid) to kill off microbes—good or bad. And a couple of hundred years ago, canning came onto the scene.†

* To make sinki in the traditional manner, radishes are cut and dried in the sun. A pit is dug in the ground and heated by fire. Radish pieces are packed tightly together in the pit and covered with leaves, boards, and mud or cow dung to create a seal. The ferment is left to proceed for at least two weeks.

† Thanks, the story goes, to a call put out by the French government. At the time, they needed a reliable way to safely preserve substantial quantities of food for Napoleon's massive army. A French brewer answered the challenge, demonstrating that food thoroughly heated in glass jars and sealed would

But far more common throughout the world and the years has been a preservation process of controlled, microbe-driven spoilage. Rather than let the wild yeasts and other microbial beasts take over as they please, we have devised ways to harness and curate the microbes rotting the food to keep it edible—and often make it even more delicious. Unlike the drive to sterilize and pasteurize our world, which has gained speed since the advent of germ theory, fermenting is an ancient acceptance, an acknowledgement—even an embracing of—the fungible, transient nature of life.

Some fruit and vegetable fermenting traditions are detailed and nuanced. But the general process of microbe-driven, salt-brine-based pickling is quite simple. Step one: Submerge a vegetable in salted liquid. Step two: Wait. Step three: Eat. No vinegar or pasteurization required. And the microbes live on in the finished product.

This is wild fermentation, a process achieved without the addition of a culture starter or backslopping. You just provide the right environment for the good guys to grow. And time. The microbes involved in this style of fermentation are supplied by the produce itself, which also serves as food for the microbes. The liquid can be added water or moisture extracted from the vegetable via salting and/or crushing. This isolates the produce from air, allowing only anaerobic microbes to thrive, thus banishing molds. The salt helps keep away harmful anaerobic microorganisms. Through their digestion process, the bacteria secrete acidic products, making the liquid too harsh for other spoiling microbes. Then some days or weeks (or in some cases, years) later, you have a safe, sour, microbe-rich food to eat. Cucumber pickle or cabbage sauerkraut, the process is pretty much the same.

But when you take a closer look inside the pickling crock, jar, or

---

remain edible over time. Not long thereafter, in the UK, the process was adapted using metal cans. Soon canned foods were a staple of armies—and then of civilians across the globe.

barrel, there is a lot more going on. And it's fascinating—as well as beneficial for us.

Just as we host an intricate, diverse world of microbes, so, too, do plants.[‡] Only a very, very small percentage of these organisms can live—and thrive—in a salty, acidic, oxygen-free environment, and these are the ones we want. It takes a little time to coax them along and build their populations, killing off the populations of undesirables, which is what makes for the dynamic process of pickling. If you've ever tried your hand at the process, you know what I mean.

You see, microbe-based pickling, to use an old saying, is not a destination so much as it is a journey. From the moment the produce is submerged in its salty brine to the time the last piece is removed, the microbial environment is undergoing constant change.

Early in the process, aerobic microbes, such as errant mold spores, will quickly be killed off when submerged away from the fresh air they need. Microbes that can't tolerate salt are also dispatched. One of the microbes that often flourishes in this stage is *Leuconostoc mesenteroides* (also found in sausage fermentation, but we'll get to that later), which is happy in a salty, anaerobic environment and puts off lactic acid, along with carbon dioxide. This is why, especially early in the pickling process, you might notice bubbles in your brine—and need to "burp" your container if you have a lid on it. Eventually, the *L. mesenteroides* does such a good job of producing lactic acid that it puts itself out of business, finding the environment too acidic for its own liking. This is when

‡ The range of microbes is wide and includes not just bacteria and yeasts, but also plant viruses, which can be hardy suckers. A fascinating side finding of one microbiome study was that some plant viruses, which don't infect humans, remain viable—and able to attack other plants—even after humans had ingested and excreted them. "You could actually reinfect a plant from the virus that was being shed out of people," UCSF's Peter Turnbaugh says. It's a microbial world after all; we just live in it.

acid-loving bacteria, such as *Lactobacillus plantarum*,[§] take the reins of the fermentation process.

But this is by no means a two-bacteria show. Several—if not dozens—of other species and strains are active throughout the process, shifting into dominance as the conditions happen to favor their preferred lifestyles. All the while, they are happily feasting on the produce and producing compounds that make the environment inhospitable to harmful microbes. The cast of characters varies based on who was present on your produce to begin with, what they are eating, and the environment they are living in. For example, a warm temperature will generally speed fermentation, whereas a cooler ambient temperature will lead to a more gradual fermentation process.

Despite all of the varied microbes involved, not all bacterial strains in our fermented foods are technically probiotic. In fact, many of them have yet to be studied thoroughly or even identified. Regardless of their probiotic status, however, they do add an additional source of microbial life and genes to our diet, possibly conferring other benefits we haven't yet uncovered.

• • • • •

Microbe-driven vegetable pickling is thought to have emerged in India some four thousand years ago, with our friend the cucumber. Since then, cucumber pickles have emerged many times in many forms across the globe, most everywhere cucumbers are cultivated. After all, one can eat only so many cucumbers in a summer. And a long winter without vegetables makes for a drab diet, as well as one that is lacking in important vitamins. Eastern European countries have a long

§ Which can be helpful in cases of IBD—and might help improve gut health overall

history of pickled cucumbers, each fermented with its own blend of spices. Cucumber pickles are found throughout Asia as well, such as the Nepalese khalpi.

These days, small-batch pickles that offer a microbe-rich brine are now starting to be sold in the refrigerated sections of more grocery stores. And the simplicity of their creation means this old style has proliferated among home fermenters as well.

• QUICK PICKLES •

Culturing-curious? Quick pickles take very little processing, are very obliging, and are easy to integrate into everyday eating. To make the iconic fermented gherkin, all you need are cucumbers, a jar, some salt and seasonings—and a little time.

Gather a pound or two of fresh cucumbers (any variety is fine), the smaller the better. Wash them and place into cleaned jars or a crock.

Add any seasonings—such as fresh dill, peeled garlic cloves, mustard seeds—to the jars. Many home picklers also suggest adding an ingredient high in tannins, such as grape leaves to help keep the pickles crunchy.

Mix a few cups of water with a tablespoon or two of salt. Pour the solution over the pickles so that they are covered. Make sure ingredients are submerged by topping them with a weight. (Many fermenters employ a brine-filled plastic bag—in case the bag springs a leak, it will be adding similarly salty water—and it also conforms to the shape of the container.) Any out-in-the-open vegetable surface welcomes yeasts to proliferate (read: mold). It's generally not harmful to the batch, but it's also not all that appealing.

Leave your pickles at room temperature, checking on them after a few days. The cucumbers should be fairly fermented in several days to a week (don't worry if the water becomes a little cloudy); wait longer if you prefer your pickles extra sour or are fermenting them in a cool location. Once the pickles are fermented to your liking, place lids on your jars and store in the fridge for weeks to a couple of months.

The Secret of Sour

The possibilities for pickling produce need not end with cucumbers. You can pickle beets, rhubarb stalks, green tomatoes, carrots, onions, garlic, papaya, ginger, daikon, turnip, kale, and of course, peppers. Just about any plant you would ever want to eat. And some you wouldn't.

One prevalent pickle is cabbage fermented into sauerkraut. I, like many people, occasionally encountered sauerkraut on my childhood plate, where it was met with general displeasure. This store-bought sauerkraut came from the depths of our pantry, usually sloshed out into a pan on the stove to be warmed before being served, limp, next to a pork chop (but mostly to just be pushed around and avoided). On New Year's Eve, my mother would coax my brothers and me into eating just one bite of the stuff for good luck, a nod to old German tradition.

And it would be good luck indeed if your microbiome would get much benefit from that type of sauerkraut. Pasteurized and cooked, it did not pack any microbes of note. Even in Germany, the homeland of sauerkraut, many people these days consume soured cabbage that is similarly lifeless.

In the traditional method of making sauerkraut, which is still practiced by many home cooks and artisanal makers, microbes abound. And for all of the potential upsides (better texture and flavor among them), the process is not difficult. Shredded cabbage is salted and massaged, releasing its moisture as cells burst. Once there is ample liquid to cover the cabbage, the shreds and their liquid are transferred to a container where the cabbage bits get pushed below the liquid. Then time and microbes are left to work their magic.

The transformation of cabbage to sauerkraut happens in stages, producing different successions of microorganisms. Like other fermented vegetables, sauerkraut contains strains of lactobacilli, which are found naturally on cabbage leaves. Recent research has revealed a surprising diversity of microbe species, including *Lactobacillus brevis*[¶] and *Lactobacillus plantarum*.[**] Helpfully, these beneficial bacterial strains also create an environment too acidic for pathogens, such as scary *Clostridium botulinum*,[††] to survive.

Beyond this basic view, science still has a rather incomplete picture of this unassuming side dish. "You'd think that we know everything, microbially speaking, about this very old vegetable fermentation," says Tufts University's Benjamin Wolfe. "It's such a simple process—cut up some cabbage, add salt, let it sit in a jar, and voilà, sauerkraut! But really basic aspects are complete mysteries." For example, he says, "are there different bacteria living on cabbages grown in different geographic regions? Do different varieties of cabbage have different types of bacteria?" At his lab outside of Boston, they are working on what he calls a "farm-to-gut sauerkraut project" because, he says, "it turns out that there is so much to learn about the humble sauerkraut."

In the meantime, in addition to microbes, sauerkraut and other fermented vegetables also add a new dimension of flavor to a meal. Sauerkraut—sour cabbage. This sour flavor comes from the acids being produced by the busy lactic-acid-forming bacteria, such as lactobacilli. Some folks find that the flavors take getting used to, but once you become accustomed to the zingy bite of fermented foods in or alongside dishes, meals will seem bland and lacking without it. And the beauty of live fermented foods (particularly those that are handmade) is that you can be somewhat in control of their flavor as well as their texture. As any experimentalist home fermenter will tell you, if you leave your ferments bubbling away for a long time, the result will be a funkier, softer

¶ Considered an immunity booster.

** Which increases antioxidants and has been shown to reduce troublesome gut bacteria.

†† Creator of the botulism toxin.

product. But steal a bite at just the right time, and you have crunchy, tangy perfection.

SAUERKRAUT

Homemade, microbe-fermented cabbage can offer a delightful crunch and a refreshingly clear acidity compared to the soggy, pasteurized, jarred variety from the store shelves. It also boasts a multitude of lactic acid bacteria that are along for the ride.

As with most fermented foods, variations on fermented cabbage popped up in many regions of the world. Some preparations use finely shredded cabbage. Others, such as Korea's kimchi, Japan's hakusai-zuke, or a variation in Eastern Europe, are made from whole fermented cabbage heads. Some are made with caraway seeds; others include seaweed or even fermented krill.

To make a traditional German-style sauerkraut, the ingredients are simple: cabbage and salt. Other seasonings, such as caraway seeds, are optional. All of the contents will go into a clean fermenting vessel: a jar, a crock, or whatever you have handy.

Chop cabbage head into thin strips and place shreds in a large bowl.

Add about a tablespoon of sea salt for every pound or two of cabbage. This draws moisture from the cabbage in addition to making the brine inhospitable to harmful microbes.

Now roll up your sleeves and get to work. With your hands, roughly smash and massage the cabbage, squeezing handfuls of it as hard as you can. The goal here is to actually break the cell walls and release as much water as possible. Don't discard the water—you'll be needing that.

If your hands are tired, take a break. And then work the shreds some more. Continue until you have extracted enough water to totally submerge your cabbage.

Now transfer the cabbage and liquid to your fermenting vessel and, if desired, add more salt and any seasonings.

Press the cabbage down below the surface of the liquid. Because some will always want to float back up (making it an enticing breeding ground for mold or kahm yeast, which forms a film across the top the surface of your fermenting liquid). So you'll want to find a way to keep it under the liquid. Some people use a plate that fits inside the vessel opening weighted down with something heavy, such as stone, a heavy dish, or a brine-filled plastic bag.

If your vessel doesn't come with a lid, use cheesecloth (or any other cloth) to cover the top to keep out dust and any more macro creatures.

Now it's time to wait. Keep the vessel in a cool space, and don't be afraid to taste throughout the process. Owing to environmental differences (particularly temperature), this is the best way to figure out when it is ready—rather than a prescribed time period. *Ready,* of course, depends on your preference. Some like it crisp and light, whereas others wait until it is soft and full of funk.

Once you're happy with the level of fermentedness, transfer all the remaining kraut and liquid into jars in the fridge to slow fermentation.

Greek Gods

Hidden here and there in classic Mediterranean cuisine, accenting sauces or adorning antipasti plates, is another gift of fermentation: cured olives. Olives are an unusual fruit to have gained such global, millennia-long popularity. For starters, right off the tree, most are far too bitter to eat. So some time several thousand years ago, people in the Mediterranean discovered that this bitterness could be fermented away.

Like standard American cucumber pickles, most olives these days are microbe-less. They are often industrially produced using lye (also known as caustic soda) to rid the olives of their bitter alkalinity. Then a simple and microbe-free brine pays lip service to the rich tradition of naturally fermented olives.

Classic natural table olives, on the other hand, still rely on a wild fermentation adventure. When you find olives made that way, they are a beautifully complex snack, complete with their complement of microbes. And to find these traditionally made olives, there is no better place to go than Greece.

When you drive northeast out of the Athenian metropolis, you can glimpse, under highway underpasses, mountains where Zeus was said to have watched fateful battles.‡‡ I take this trip to find microbe-fermented olives packed into a small car with a cigarette-smoking Greek microbiologist and a scientist colleague. After about an hour's drive from central Athens, we cross a bridge onto Evia (also known as Euboea), Greece's second-largest island. Here hills blanketed in old olive groves rise up from the sea and then ascend farther into dramatic, pine-covered mountains. Leaving the main thoroughfares and heading across the island, we pass through small towns, wait for grazing goats crossing the street, and see old men collecting wild greens on the side of the road.

Off one of these side roads is a small coastal village called Rovies. Here we find the man we have come to see, Nicos Vallis. He has called this stretch of land home for some thirty-five years. Although past retirement age, he still works the olive orchards that have been owned by his wife's family since the Greeks gained the land back from the Turks in the 1820s. He is also president of the local olive cooperative and is working to turn local orchards organic—and to turn the cooperative's processing factory from lye processing back to natural fermentation. Today the factory processes about 100,000 tons of olives a year. But when Vallis, who was born and educated in Alexandria, Egypt, moved to this picturesque place in the early 1980s, he didn't know a thing about growing olives—and he knew even less about fermenting them.

The cooperative's small factory is perched just above the shore. It

‡‡ According to legend, the gift of the olive tree is what earned the goddess Athena the honor of being the namesake of Athens. This olive tree, planted at the Acropolis, was said to have lived for centuries after the founding of the city.

has a view down to the gulf in one direction and up to the towering hillsides in the other. The quiet sea is just an olive pit's throw away. Sun warms outdoor fermenting tanks, and the light breeze carries the song of chirping birds against the backdrop of the mountains. Despite the occasional mechanical noise from sorting conveyor belts inside the factory and an ongoing conversation about microbial genes, the endeavor has a distinctly timeless feel.

Inside a long building, clerestory windows illuminate the Spartan space. The vaulted ceiling lends an air of devotion to the ritual happening below. Here olives will take eight to nine months to transform from inedibly bitter fruits to savory snacks. Rows of deep rectangular basins are lined up under a second-story walkway. These are the original vats, some still in use, where the olives slowly ferment with salt, water, and just a touch of lactic acid to jump-start the fermentation. Old wooden planks sealed with paraffin wax help keep the air and debris out. Wooden rakes on long handles wait at the ready to stir batches in an enduring practice.

Vallis takes pride in his long-brined olives and the simplicity of the process: "We just put the olives in the brine," he says. In comparison to olives that have been treated with lye and cured in just a few months, his jarred green olives are darker, a bit more muted, and take several months to mature. They are crunchier and bitterer. Unfortunately, he says, customers these days are used to the vivid color, softer texture, and duller flavor of lye-treated olives. Sampling olives straight off the production line, I find that they have a powerful complex flavor with competing salty and spicy notes. Growing up with watery black olives (such a favorite that I was delighted to receive a can of them in my Christmas stocking each year as a child), and later eating oiled-and-herbed olives at self-important restaurants, I had never really thought of olives as a pickled product. But these Greek olives divulge their elaborate preparation in a riot of sophisticated flavor.

To better understand this process, Vallis has been partnering with my driving companion, microbiologist Effie Tsakalidou, and her research lab at the Agricultural University of Athens.

Naturally fermented olives generally pass through three phases. In the first couple of days, when the pH is still neutral, a variety of microbes thrive. Next, as the pH dips, lactic acid bacteria, such as *Lactobacillus plantarum*, and acid-tolerant yeasts take over, lowering the pH further over the next couple of weeks. These populations continue into the final phase of fermentation, with the flourishing of yeasts such as *Pichia anomala* (also active in the early fermentation stages of wine), *Saccharomyces chevalieri* (another wine yeast), and others.

Broad surveys of various spontaneously fermented olives have found rich microbial communities—and some potentially probiotic strains. One team that looked at traditionally fermented table olives made in Sicily found that just three and a half ounces of these olives contained some 1 billion live cells of *L. plantarum* or *L. paracasei* (both of which have probiotic strains). The authors of another paper found 238 different strains of various lactic acid bacteria. Among those, 17 showed promise as potential probiotics, with one strain in particular, *L. plantarum* (strain S11T3E), revealing the best probiotic properties, inhibiting harmful microbes such as listeria.

Despite his decades of cultivating and culturing olives, Vallis seems more passionate than ever about spreading the gospel of naturally fermented olives. "I am just trying to make it known to as many people as I can," he says. His brother-in-law Stefano is of the mind that microbe-fermented olives simply speak for themselves: "You don't need to prove anything else," he says. "From the taste, you can tell." And after sampling them, I am inclined to agree.

Traditional Transformations

On the other side of the globe from Greece's slow-fermented olives, another country has been maintaining a lively and nuanced tradition of pickling for thousands of years. Traditional Japanese cuisine is brimming with pickles—from fast-fermented lotus root to slow-pickled plums.

The umbrella term for pickled food in Japanese is *tsukemono* (something to the effect of "pickled things"). Throughout the cuisine, the word *zuke* can be added to particular foods or processes to indicate "pickle." Beyond simple souring by microbes, tsukemono is a larger way of thinking about food. "Tsukemono is transformation," says Elizabeth Andoh, an American food writer and cook who has been living in Japan for more than fifty years. She gives me her perspectives on Japanese pickling when we meet for tea at a hotel by Shibuya Crossing.

Andoh hosts workshops on tsukemono in her Tokyo home, from where, on a clear day, you can glimpse the revered Mount Fuji. Before the start of each pickling program she asks participants (often from the United States or the UK) what they consider the essence and purpose of pickles to be. Many say something about vinegar, she notes, "and almost all of them talk about summer bounty being put up for the winter. When you say *pickles* to an American—or at least an English-speaking—audience, that's usually what they expect. And tsukemono's none of that," she says. "Yes, some of it is about putting aside for the winter." But mostly, "it's year-round." It is an expression of seasons and of tiny micro-regions and of heritage, she says, rattling off names of various zuke and prefectures in a cascade of beautifully rounded syllables. These fermentation products adorn and accent almost every meal as a tangy companion and purported health aid.

One zuke that seems ever-present in Japanese meals—from sushi spreads to bento lunch boxes—is takuan, pickled daikon radish, named for the influential seventeenth-century Buddhist writer Takuan Soho, who is credited with creating it. The pale ocher root turns bright yellow when pickled, thanks to the addition of gardenia pods, and is often eaten by the slice at the end of a meal. The ones that I sampled, rather than a spicy bite or crunch, yielded with a soft and mellow mouthfeel. But the addition of extra chilis or different fermentation lengths can yield plenty of spice or a firm slice in other preparations.

Another seemingly omnipresent pickled food in Japan is the umeboshi plum, made from ume, a fruit somewhere between a conventional plum and an apricot. These soft pickled fruits are often eaten with breakfast or lunch, usually along with rice. In bento boxes, they are often placed in the center of a rectangular bed of white rice, creating what is known as hinomaru bento, after the name for the Japanese flag it resembles. Umeboshi are said to promote stamina[§§] and ward off illness, thanks to their pH-lowering powers. One popular story asserts these powerful pickles even prevented an outbreak of foodborne illness from contaminated rice in bento boxes. This feat, Andoh says, "demonstrates the wisdom of the ancients."

The tradition of umeboshi goes back at least several hundred years, with early records suggesting the origin of umeboshi more than a thousand years ago. And in many places, the process of making them hasn't changed much in those intervening generations. Typically, the sour ume are picked in June, sun-dried, salted, packed tightly with red shiso leaves, and then fermented to create the prized product. Andoh notes that to start the process, you need a sizable volume of plums to create the correct weight and mass for packing and pickling. In fact, she says, "it's almost impossible to produce decent umeboshi with less than ten kilos" of the fruit—and that's an absolute minimum, she notes. She used to be part of a group that would come together each spring to pickle

§§ If they were good enough for samurai going into battle, they ought to be good enough for the rest of us.

umeboshi—about thirty kilograms, or sixty-six pounds—and then divide the product up among the members. Andoh would take a few pounds back with her to Tokyo, "which was just about a year's supply for me." But, she says, "most Japanese will store the finished plums for several years before eating. Much like aged wines, two- or three-year plums are usually preferred to those freshly made. They mellow over a period of years—they do not spoil." If prepared right, Andoh says, umeboshi could "theoretically keep forever."

Umeboshi's flavor can be quite a shock to the unaccustomed palate. They are potently sour and quite salty. Even some in Japan find umeboshi to be a bit overwhelming. One group of Japanese researchers looking to study the activation of taste centers in the brain asked Japanese volunteers to just *think* about eating these plums as they underwent an fMRI scan. It was with this conjuring of a "strong and sour taste" alone, the researchers wrote, that they were able to "observe cerebral activation patterns"—without volunteers so much as putting a pickled plum to their lips.

• • • • •

In addition to these more familiar types of brine-based pickling, Japan is also home to a very particular form of pickling, one that relies not on a liquid salt bath to control the fermentation but instead on a carefully cultivated bed of damp rice bran. This is the hidden pickling world of nuka.

Nukazuke are time-sensitive on a different scale than traditionally submerged pickles. Most can be ready in a day—or in as little as a few hours, Andoh says. These "quick pickles" are no less microbially fascinating than longer ferments. In fact, their process can make them even more intriguing.

To start a nuka bed, which is kept in a dedicated nuka pot, toasted rice bran is mixed with water and salt. This creates a damp mush in which fruits and vegetables can be transformed. Some people also add kelp or beer, along with raw vegetables for flavor and for extra microbial inoculation. The blend is mixed by hand once or twice a day, and new vegetables or peels are added for more fresh microbes. After a week or two, it is dense with yeasts and lactobacilli—and is ready to start pickling.

This method of pickling might have emerged when rice was first being milled, separating the white center from the outer bran (a process already in practice in Japan by the tenth century). And it remains an additional way of putting this scrap material to productive use. But in addition to this conservation ethos, which runs deep through Japanese culture and cuisine, nuka pickles are also a celebration of the seasons.

The seasons affect the nuka in numerous ways. To start, there is the temperature, with higher temperatures yielding shorter pickling times. In the summer, for example, Andoh says, she will place cucumbers in the bran mash "just before lunch to share for dinner. In the winter, it's more like twenty-four hours. So after dinner, I'll put it in for the next night."

The other way the nuka changes with the seasons is in what produce gets placed into the pickling mash. "There are certain things that are pickled at certain times of the year," Andoh says. In the summer, for example, eggplants are plentiful. They are often rubbed with salt and alum before being placed in the nuka bed. After a summer of these ferments, however, "the nuka bed gets a bit musty," Andoh says. Just when the bed needs rescuing, persimmons arrive into season. In the fall, "their peels are put in the nuka pot to extract undesired compounds [from the bran mash] and to add a bit of sweetness. I'm sure that there are chemical formulas, but it was cumulative kitchen wisdom. Experience along the way," she says.[55]

[55] Most of the adjustments were historically done simply by knowing one's nuka. "My mother-in-law knew that if she put a clove of garlic in the pickle pot and it turned blue, and a nob of ginger in and it turned pink, that the pH was just right," Andoh says. "She didn't know why, but she knew it was just right. When she had to tinker with it, and she wasn't quite sure it was ready yet, that's how she figured it out. It's all observation."

Which leads to the question of what exactly is going on inside this crock of damp bran mush.

That inquiry might ultimately be unanswerable—at least in the specific. For starters, each household's nuka bed, also known as nukadoko, lives in its own environment, open frequently to the air (unlike the somewhat more protected anaerobic environment of a brine liquid). It is also mixed daily by bare hands. This means that each nuka bed is different from the next: a personal blend of microbes from the environment and the pickler's own hands. The characteristics of the bed, as Andoh alluded to, also change with the foods buried in the bed, each bringing with it its own combination of bacteria and yeasts.

Because of all of these factors, there is no essential singular nuka microbe profile. Most pots are replete with lactobacilli species. In one nuka bed, a team of researchers found a hardy strain of *Enterococcus faecium* (some strains of which are used as probiotics) that was effective at fending off pathogens, such as listeria. It is a rich and dynamic environment, ripe for further study.

For all of their benefits, nuka pots don't just get started on a whim. These are living beds, some of which are generations old and cared for continuously. Without regular mixing and attention, a nuka bed can die—and with it, its unique properties and microbial life. People regard their nuka pots almost as a pet. And this pickling method is still widely practiced in home kitchens throughout the country.

Andoh relishes the rhythm of nuka pickles and says her nuka pot "is very precious to me"—although it is not, she notes, entirely hers. "My mother-in-law's mother-in-law started it." She estimates that it has been going for about one hundred and fifty years. The practice of pickling in Japan encapsulates "that transformation of knowledge"—over both time and space, Andoh says.

King Kimchi

Making traditional pickles, sauerkrauts, and olives are not the only ways people have found to turn vegetables into delicious, live fermented foods. Seoul's busy Gyeongdong Market is a showcase of many of the other techniques.

Daniel Gray, an American-educated South Korean restaurateur and food expert, and my market guide for the day, points out the incredible variety of fermented products, wildly different from stall to stall. In addition to pickled vegetables, the market is also full of fermented sauces, live vinegars, bean pastes, crustaceans, and more.

Gray notes that there has long been the sense that these foods are good for you. "Koreans think of food as medicine," he says. "Taste is very important, but if it's healthy, it kind of usurps everything else." Even beyond that, he says, "food makes you who you are." A notion that would be good for many of us to pause to consider more often in our daily lives.

This tie to tradition also defines the process. Fermentation recipes are handed down through generations, he says. You might try a tweak here and there, but the mark of a successful fermentation is if it "reminds you of your mother's recipe," he says.

In this wild landscape of ancient pickling traditions, there is one style, however, that reigns: kimchi.

Technically fermented cabbage, it is so much more than just Korean sauerkraut. In its most standard form, it also includes green onions, radish, chili, ginger, salt, and fermented fish sauce, all packed together and left to mature for months in a clay pot buried in the ground.

Kimchi has long been prized for its nutritional qualities, offering generous amounts of vitamins and fiber. And when made traditionally, it also contains loads of the aptly named *Lactobacillus kimchii* cultures.

But it is not always love at first bite.

In Korea, kimchi is everywhere. It comes as a side, in savory pancakes, and as the base of soups and stews. It even makes an appearance at a hotel breakfast buffet. Sitting down at Korean restaurants in the States, I would pick dutifully at the bright-red-specked cabbage-and-vegetable kimchi in the small dish—using bites of rice as a chaser. I am not a particularly picky eater (I count ants, Rocky Mountain oysters, and live octopus arms among my gustatory conquests—not terribly minding any of them), and I grew up under the impression that pickled jalapeños were a standard hamburger topping. But somehow to me kimchi was more challenging, with a spicy funk that resisted acclimation.

Kimchi is a rich and fascinating food, full of complex, shifting microbial communities, as well as important prebiotic fibers. With such an impressive résumé, the food deserves more than a few trepid tastes. Even if a bite of it my first morning in the country made the coming week unfurl before me like an endless leaf of sharp-flavored napa cabbage. Still, I needed to learn more about this ubiquitous, multifarious, and microbe-filled food.

• • • • •

Kimchi is unquestionably South Korea's national dish. And that's a designation Koreans don't take lightly. They consume an average of forty pounds of kimchi per person each year.*** The country has a beautifully designed museum dedicated entirely to kimchi. And the first Korean astronaut, Yi So-yeon, took kimchi with her into space.

*** If we want to put that in microbe terms, that's something like an annual dose of 1,814,370,000,000 lactic acid bacteria—with many other bugs to boot—flowing into each person's intestinal tract. Put in the context of our guts, that's a boost equivalent to about one-twentieth of the body's native microbe population getting introduced to the body each year through kimchi alone, interacting with the body's systems along the way. And if forty pounds of kimchi sounds like an impossible feat, broken down over the year, it's actually less than two ounces a day. A reminder that these fermented foods are not the main attraction, but a carefully integrated aspect of the overall cuisine.

Kimchi is so central to Korean culture that the process has its own name, *kimjang* (or *gimjang*). Kimjang Day is still celebrated in many villages and has been granted UNESCO Intangible Cultural Heritage status. On this day each fall, women come together, often in a large open-air courtyard, to turn hundreds of heads of cabbage into the coming year's supply of kimchi.

The basic process of making cabbage kimchi starts with napa cabbage. Soak the whole cabbage in a saltwater bath. Separately, mix chives, radishes, garlic, ginger, hot red pepper, and fermented shrimp. Now add the spice mixture to the cabbage, rubbing it into each layer of leaves. Tuck the bundle into a container, ferment, and then enjoy.

Despite the prevalence of this type of "standard" kimchi, the food is as diverse as the country itself—with local recipes being handed down over generations in mountainside towns and seaside villages. There are more than 180 distinct recognized types of Korean kimchi.

Some locales make a radish-based kimchi, which has just a small amount of cabbage. Others use fruit to create a sweet-and-sour blend, such as royal kimchi pear, made with radishes, pears, citrus, tinted red by pomegranate and red pepper powder, and arranged and sliced into beautiful florets. There is also a white cabbage kimchi. For this one, the filling contains chives, radish, garlic, jujubes, ginger, and shredded red pepper. The soaked napa cabbages are stuffed with these seasonings before fermented fish brine is poured over the whole thing for fermenting.

Other kimchis feature not vegetables but fish as a central filling. One area of the country creates a green onion and dried squid kimchi. For this version, squid is first stir-fried so that it retains its chewy texture even after fermentation. Other locales prefer their fish kimchi to be smoother. For a cutlassfish-based kimchi, fish are first fermented separately long enough so that their bones disintegrate. The result is a tender filling in a much-prized dish. "We bury cutlassfish kimchi underground, as if it were a secret treasure box," says one older Korean woman, who still makes this type in the traditional style, in a short Museum Kimchikan documentary about the process. This kimchi is considered a delicacy, served only for special occasions and during visits from important guests.

All of the wild varieties of kimchi are just the start to this food's fascinating diversity. The varied microbes involved in kimchi ferments come from the ingredients—often including those that have already been fermented, such as fish sauce. And come they do. A single gram of fermented kimchi contains some 100 million lactic acid bacteria.

• • • • •

To learn more about this lively, pungent food from the scientific perspective, one rainy spring morning I trekked out to the Korea Food Research Institute. The institute is a large building perched on a wooded hillside a ways outside of Seoul (a trip that required navigating the city's bus system, with the help of a good-natured interpreter). A soggy walk up the hill led to the institute's gatehouse, and after we gained entrance, we set off through the corridors to find Myung-Ki Lee, a lead microbiological researcher there, who has spent years studying fermented foods.

Now if we thought we were just beginning to understand the microbial dynamics of sauerkraut, kimchi is a whole other can of worms (or bacteria, as it were). A single variety of kimchi might contain more than one hundred different types of microorganisms. Such a diverse array of microbes is even more mind-boggling when you consider that in practice, "for a meal in winter, Koreans traditionally had three different types of kimchi on the table," says Lee, speaking partly in Korean and partly in English, both filtered through the interpreter. Despite grappling with these dauntingly large and diverse populations of microbes, he and his colleagues are intent on better understanding the complex microscopic world of kimchi.

Kimchi is a rich and fascinating food for a microbiologist in part because it contains so many different microenvironments. Uniform fermented foods such as yogurt generally have a fairly plain substrate—one area of a yogurt container is much like the others.††† But in kimchi, there is a veritable planet of land and sea habitats for various microbes to find their niches. "Kimchi has leaves, it has roots, it has solid ecosystems, it has liquid ecosystems," Lee says. "Kimchi, because it has different ecosystems, it can support many types of microorganisms."

One study of five different commercially available cabbage kimchis found 348 separate strains of microbes.‡‡‡ The most common microorganism that study found was *Weissella koreensis*, a lactic acid bacteria strain that has antimicrobial properties and might help fight obesity. The researchers also found strains of bacteria less common in other fermented foods, including *Lactobacillus sakei*, which is also in some aged sausages and has been shown to help modulate the immune system and to relieve eczema. Additionally, the researchers came across at least one entirely new *Lactobacillus* species that had not previously been described, suggesting that with all of the homegrown kimchis out there, there is surely a world of species and strains waiting to be discovered.

And these impressive snapshots tell only part of the story. Kimchi is eaten throughout the fermentation process. "Once we make kimchi, we consume it from the start," Lee says. "And we can continue to consume it as time goes by."

Because the ferment is constantly progressing, Lee says, "the dominant microorganisms tend to change," making it a rich source of various microbes—but an exceedingly difficult one to characterize. "Lactic acid bacteria tend to be the dominant microorganisms," he notes. But at

††† Except in rare instances, such as viili, in which a yeast grows on the surface, helping itself to the oxygen above and the nutrients below.

‡‡‡ Compare that with the handful of strains present in most probiotic supplements or in a store-bought yogurt.

different phases in the fermentation succession, an entirely different genus might come to prominence. "In the initial stage, *Leuconostoc* tends to be dominant. But in later stages, *Lactobacillus* tends to be dominant," he says. In this sense, he explains, it's quite different from cheese and yogurt, which have a finished product that is relatively standard, microbially speaking. When I ask Lee if he has ever considered sequencing the "kimchi microbiome," he just laughs. There is of course no single profile to sequence.

These battling microbial populations also affect kimchi's texture and flavor in nuanced ways over the duration of its dynamic life. As the ferment progresses, vegetables that were originally crunchy become soft. Flavors ripen from fresh to funky.

To gain a better hold on the macro outcomes of this process for the kimchi consumer, manufacturers have developed dedicated kimchi refrigerators. Resembling a dorm fridge or a wine chiller in size, these small appliances play an outsize role in the country's continued kimchi consumption. Part of it is logistical with a scarcity of undisturbed ground around Seoul in which to bury your kimchi pot, and kimchi is a rather copious, long-term, and potent thing to store with your daily foods.

As Lee notes, these refrigerators also have another important feature that sets them apart: fine temperature controls. "The most important factor in fermentation is the control of temperature," he says. "As the temperature changes, different types of microorganisms can flourish." Lee actually worked with producers of one kimchi fridge to dial in temperature settings so that consumers could ferment their kimchi just to their liking. "So, for example, if you want to have more carbonated taste to the kimchi, you can adjust the temperature, and in that way achieve that goal," he says. "Or if you want the kimchi to be strong in sour taste, then you just adjust the temperature so that the microorganisms that make the kimchi more sour can flourish." It is a fascinating intersection of tradition and technology.

Despite its prevalence and popularity and microbial richness, kimchi has been surprisingly difficult to classify as a categorically "probiotic" food, precisely because it is so diverse. "That's the exact problem," Lee says, "because there are so many kinds of kimchi." His work has focused on classic cabbage kimchi. But "there are many different types of side ingredients that are added to kimchi"—which change the microbial dynamics—"so it's very difficult to determine which functional effect comes from which kind of cabbage, and which comes from the side ingredients." And very few of those organisms have been studied and tested to discover what specific benefits they might confer upon the human consumer. But with numerous *Lactobacillus* species, including the immune-booster *L. sakei*, as well as species from many other genera, the potential microbe-modulated health benefits of kimchi are likely as numerous and varied as the dish itself.

Kimchi isn't the only fermented food in the traditional Korean repertoire. But, Lee says, "the best fermented food for people to eat is kimchi. For the living organisms." He laments, though, that people tend not to eat quite as many types of kimchi as they once did, eating only one or two types at a meal. He contends that "it's very important to have three kinds because it ensures the diversity of microorganisms. The diversity of microorganisms is very important"—for the kimchi and for our health, he says.

In addition to the varied world of microbes, kimchi also has something else to offer the gut: fibers to feed resident microbes. Inulin from garlic or onions, for example, make it a natural synbiotic, a single food that provides probiotics and relevant prebiotics. Some newer kimchi products are actually ensuring this status by adding oligosaccharides for a marketable prebiotic component. But no matter what kind you find or make, it is one of the most microbe-rich, prebiotic-filled fermented foods out there. Even if some folks find its flavors take a little getting used to.

• KIMCHI •

Although seemingly a culinary world away from sauerkraut, kimchi is made from the same basic ingredients: cabbage, salt, and, time.

To prepare the classic kimchi I learned to make at the Kimchi Academy in Seoul, you will need napa cabbage, radish, green onion, the Korean red chili pepper powder called gochugaru, sesame seeds, salt, fish sauce, and ginger paste (which can be bought at a specialty shop or made at home by pureeing fresh gingerroot with a little bit of water). Optional: sugar, shrimp paste, glutinous rice flour (turned into a paste by adding water). Additionally, a bowl for soaking the cabbage and a vessel for fermenting it. This is for one head of cabbage; scale up as desired.

Soak a whole head of napa cabbage in salted water.

Meanwhile, julienne a small daikon radish.

Slice two green onions into four pieces each.

Drain the soaked cabbage and rinse lightly. Rub salt between all of the cabbage leaves

In a small bowl blend:

3 TABLESPOONS RED CHILI PEPPER POWDER
1 TABLESPOON SESAME SEEDS
1 TABLESPOON SUGAR (OPTIONAL, FOR FASTER FERMENTATION)
1 TABLESPOON FISH SAUCE
1 TABLESPOON SHRIMP PASTE (OPTIONAL)
1 TABLESPOON GLUTINOUS RICE FLOUR PASTE (OPTIONAL)
1 TABLESPOON GINGER PASTE

Blend the mixture with your hands, then mix in the radish and green onion.

Use your hands to spread the mixture between each of the leaves of the cabbage, smearing the outside as well.

Fold the head of cabbage in half, and use the long outer leaves to wrap it into a neat package. Nestle it into a fermentation vessel and make sure it is below the surface of the liquid that squeezes out of the cabbage heads, using a weight to keep it down.

Traditionally, dozens of these bundles would be tucked together in a large crock known as an onggi, submerging the cabbage bundles in their own juices. The onggi would then be buried in the ground for a cool and steady temperature and humidity. The crock's neck would be left above ground for easy access to the stores throughout the winter. But the kimchi can also be fermented in any jar or pot in a cool location. Just be sure to keep the cabbage submerged as you would for sauerkraut to prevent yeasts from alighting atop the contents.

Plant Superpowers

Now that we've dived into the world of the pickle, let's see how else these foods can benefit our guts—this time feeding our microbes as well.

Perhaps it's no surprise that the foods such as cabbage, onion, and garlic that pop up in pickling recipes can feed not just pickling microbes but also the microbes that live in our guts. In addition to being excellent pickling subjects, many vegetables (and some fruits) are among the best sources of prebiotic fibers we can get in our diets. For example, inulin and fructooligosaccharides (FOS) are some of the most common prebiotics to feed our resident microbes, and they are both found throughout the plant kingdom—perhaps yet another reason diets heavy on plant consumption are linked to lower rates of preventable diseases.

In the famously healthful Mediterranean diet, for example, plant-based foods take center stage. In many Greek traditions, meat has been an occasional extra, a once-a-week indulgence. The rest of the time, plates are filled with vegetables, beans, grains, cheese, and of course, yogurt. It might have been born of frugality and harsh conditions. But it might also have been serving the Greeks—and their microbes—well over the generations.

In Greece, the produce on plates often still reflects the bounty of the region and the season. And nowhere is this better reflected than in the freshly gathered horta. As much a ritual as an ingredient, horta is a catchall term for the wild-foraged greens that find their way into a myriad of dishes—or are served on their own, steamed, with olive oil and a squeeze of lemon, or as a salad. Horta—casually translated into English as weeds—can include dandelion greens, purslane, mustard greens, wild fennel, stinging nettles (cooked), and hundreds of other wild-foraged plants, the variety depending on the season, terrain, and weather. Because these gathered horta include such a wide range of foods, they provide a diverse complement of nutrients and fibers. "We have a lot of weeds in our diet, so we have a lot of fibers," says Kostas Papadimitriou, a microbiologist at the Agricultural University of Athens.[§§§] A typical taverna lunch I enjoyed in a perilously steep Greek mountainside village included wilted horta, a fava bean puree with onions, feta cheese pie, table olives, and smoked fish served in oil. It was a rich and filling lunch, containing live microbes and a range of prebiotic, microbe-feeding compounds.

Now, the image of Greek grandmothers climbing into the hills with their burlap sacks and a knife might seem a quaint one of a history long gone. But gathering horta is still very much a common practice for both old and young—even for city-dwelling Athenians. For many, it is a practice they associate with the simpler and more healthful village life of their childhoods or parents or grandparents. They will make a special trip outside of town to nearby mountains to collect horta. And many women sell their own hillside-foraged goods at the markets in town for those who can't get away. Horta are even for sale at the chain supermarkets. Picking one's own horta, though, is seen as a superior option for

§§§ "We don't eat only salads of lettuces," Papadimitriou notes. He emphasizes the wide range of what Greeks might think of a salad—which could involve horta, vegetables, wheat, beans, and more.

reasons of economy as well as for getting the pleasure and health benefits of exercise and being out in the fresh air. These benefits alone would probably do most of us—and our microbes—quite a bit of good as well.

After a delicious horta and feta pie served at my seaside pension, I inquired about the contents and was told: "If you ask the cook, who also gathered the horta,¶¶¶ how many species were in the pie, she would say, 'more than a hundred!' " Whether this is exactly right or not, clearly the notion of serving a wide range of plants was a point of pride and tradition.

A similar mind-set is behind a tourlou, which was defined for me by my Greek hosts as: "everything mixed up together"—whether clothing ("whatever you could find in your closet") or food ("whatever you find in your garden"). The version I enjoyed was a sort of Greek version of ratatouille, a mixed veggie stew, surely with some horta mixed in, served with a yogurt topping.

With food, as in life, Papadimitriou says, there is an old Greek saying: *pan metron ariston*, "which is 'you have to have a balance in everything in your life.' "

¶¶¶ Who, of course, happens to be the innkeepers' mother.

• GREEK HORTA AND CHICKPEA SALAD •

Greek cuisine can be beautiful for its simplicity and its purity of ingredients. This is a recipe for a simple salad inspired by my meals in the Greek countryside and a prebiotic-filled recipe from the Stanford scientists Justin and Erica Sonnenburg's book *The Good Gut.* This dish works as a side or can be served on its own for a filling lunch. The ingredients include chickpeas (just half a cup of which has more than a third of the daily recommended amount of fiber), horta (no foraging required: you can use dandelion, mustard and/or other greens from the market), bell pepper (any color, depending on your preference), red onion, lemon juice, olive oil, feta cheese, and salt and pepper.

Cook chickpeas or use drained canned. Place about four cups in a large bowl.

Slightly wilt a bunch or two of greens (the more variety, the better) with a little water on the stovetop or in the microwave.

Core and chop one or two bell peppers.

Finely slice half a red onion.

Combine the vegetables in a large bowl with the chickpeas and stir to mix.

Squeeze lemon juice and drizzle olive oil over the top. Toss to coat.

Top with feta cheese and salt and pepper to taste.

A focus on food and wellness is of course not limited to the Mediterranean. Japan is known as well for its traditionally plant-centric diet, an array of vegetables lending a diversity of fibers to nearly every meal. In Tokyo, residents can visit one of the city's many regional antenna shops, which stock the specialties of specific provinces. In the Yamagata antenna shop in the springtime, for example, I spot bunches of wild-foraged greens and fiddlehead ferns. The Okinawa shop, which maintains a Polynesian feel, offers up a spread of fabulous produce, from bitter gourd (another name for bitter melon), to fresh turmeric root, to burdock.

The beauty and benefit of many Japanese meals is simply in the multitude of different foods. Not every meal includes dozens of minute dishes. But most include several different items, served in small portions. One late morning, I join Midwestern-born Japanese American chef, sommelier, and author Yukari Sakamoto for lunch at a lovely rice-shop-cum-restaurant in Tokyo's tony Ginza neighborhood. There, a standard set lunch comes with ten separate small dishes. These include a miso broth, daikon pickles, braised greens, a shrimp cake with fava beans, and numerous other vegetables. And, of course, rice. This, says Sakamoto, is how the Japanese maintain a healthy diet: lots of different things in small amounts, helping to ensure a variety of nutrients, bacteria, and prebiotic fibers.

Despite Japan's reputation as a healthy and long-lived society, the prevalence of so-called Western diseases, such as inflammatory bowel disease, has skyrocketed there. The shift may have a lot to do with rapid dietary changes, scientists suspect. The island of Japan was for more than two centuries also a cultural island, largely sealed off from the rest of the world until about a century and a half ago. Still, at that point, writes a team of Japanese medical researchers, "only a small portion of Japanese people could afford Western food, while the vast majority continued to eat frugal Japanese foods for an additional 100 years. A typical Japanese diet at that time was a simple vegetarian meal composed of unthreshed rice mixed with barley, miso soup with root vegetables and/or tofu, small grilled fermented fish, and fermented pickled vegetables." After the end of World War II, as the country rebuilt and gained more widespread prosperity, documents show a "rapid increased intake of sugar-rich carbonated beverages, fat- and carbohydrate-rich Western snacks (e.g., potato chips), and animal protein and fat, and a concurrent rapid decrease in the intake of dietary fiber," they note. Alongside these changes was the perhaps predictable growth in gut-based diseases, such as ulcerative colitis and Crohn's disease. The researchers discovered, however, that they could help many of their current patients with IBD by slowly reintroducing prebiotic-heavy whole foods (such as barley mixed with rice, seaweed, and fruits and vegetables). "Asian societies are at a crossroads between a Western-style and a traditional high-fiber, low-fat, and fermenter-rich diet," they conclude. "Clinicians should encourage these traditional foods to promote public welfare." The rest of us might do well to take that advice, too.

• • • • •

In China, there is also a long history of using diet to promote and restore health. And much of that starts in the gut. "Actually, we don't have a very clear distinction between food and medicine in China traditionally," says Liping Zhao, a microbiologist at Shanghai Jiao Tong University, when I meet him at the Keystone Symposia on Molecular and Cellular Biology's meeting on the gut microbiota and host physiology. Many plants are used both as food and as medicine, reinforcing "the concept of food as medicine," he says. "I tell people that if you eat your food as if it is medicine, you don't need to eat medicine as food."****

One way to do this is to search out how previous generations ate. "If you look at the traditional diet, traditionally people eat food as a whole entity," Zhao says. "You don't process, you don't refine," leaving nutrients and fibers intact for the consumer—and their microbes. A basic diet is rich in complex carbohydrates, says Zhao. But he notes that today, even in rural areas in China, it is getting more and more difficult to find a truly traditional diet. "We used to have a plant-based diet. But I think it's almost the same everywhere—when people are getting more money,

**** An echo of the famed quote frequently attributed to Greek physician Hippocrates: "Let food be thy medicine and medicine be thy food." (The veracity of this quote and its source, however, has more recently been called into question.)

they immediately increase the animal food in their diet. That dramatically changes their dietary structure and also their health." He laments that although such economic development can be a boon in many ways, "it's really sad to see that many of the traditional, supposedly healthy diets are actually fading, are disappearing." This decline started on a massive scale with growing economic prosperity and when many habits of old were upturned, he notes. "China probably has conducted the biggest trial in world history. We changed the diet of one billion people over twenty years. Then we changed the disease spectrum almost overnight," he says.

Chinese researchers are now looking back to the long history of medicinal foods for clues about better nourishing the microbiome in the future. One clinical trial that gave patients with type 2 diabetes traditional herbs found that consuming these plants boosted *Faecalibacterium prausnitzii* in the gut, a microbe known to produce anti-inflammatory compounds. In another trial, morbidly obese children were given a strict diet high in plant-based complex carbohydrates. Not only did many of them see their weight decrease, but there were also marked shifts in their microbiotas—and the compounds those microbes were producing.

One of the key foods used in this latter trial was bitter melon. These are long, bumpy green gourds that have historically been used to treat intestinal tract ailments and also crop up again and again in longevity-linked diets. "In China, there are many foods which may be sustaining some beneficial bacteria," Zhao says. "The one we study the most is bitter melon," which has roots in ancient Chinese medicine. In that tradition, it is thought that "bitterness can cleanse fire from your body," he notes. "If you regard fire as inflammation, bitter foods may actually reduce the inflammation." But for a long time it wasn't clear how this actually worked, particularly with compounds from the food skipping through most of the human digestive tract. Indeed, "the 'drug' is largely remaining in the intestine, doesn't get into the bloodstream much," he says. "Based on Western pharmacology, if a compound does not get in the bloodstream, it should not work. This paradox actually can be resolved if you look at the gut microbiota as a target." Our microbes are digesting these otherwise indigestible compounds into other beneficial compounds that the body can absorb. Which is all the more reason to remember the gut microbiome as a powerful mediator between food and overall health.

• • • • •

Fruits and vegetables are, of course, not the only plants that have a beneficial impact on the microbiome. And there is another fermentation tradition that predates even fruit and vegetable ferments. In many forms, it brings us escape. In some forms, it also brings us a rich microbial dram.

CHAPTER SIX

Intoxicating Ferments
Grains

The earliest ferments were probably not batches of sauerkraut or kimchi, sprung from unattended cabbage. Nor were they likely the yogurts or kefirs, discovered after a long journey on horseback. Humans likely got an early taste for ferments from alcohol.

Fermented grain products range from drinks inoculated with mouth microbes to a finely molded rice that is a staple in many Asian fermentation traditions. They are not just your average ale. Certainly fermented grains don't end with beverages. But that is where our journey begins.

Cheers

When microbes meet grain, they can digest its carbohydrates, creating alcohol. These ferments are often driven by yeasts, which off-gas carbon dioxide in the process, creating carbonation. From wild-fermented beers to traditional sake, these beverages are not only culturally important but also, when created in the traditional way, they are also rich in flavors, intoxicating compounds, and yes, sometimes even some unexpected microbes.

Today, few alcoholic beverages contain live microbes when they reach your cup. But if you take time to search them out, there are still some that have a rich microbial collection to offer.

The process for making the earliest, microbe-rich drinks would probably give modern-day brewers and distillers—not to mention their customers—pause. Rather than using a sprinkling of yeasts to jump-start the fermentation of mashed grains, the early creators of these drinks likely started the process with mastication.

• • • • •

Grains are rather tough for most microbes to start breaking down without help. So humans found a way around that by chewing them up first. Using a pre-chewed starter has a double benefit of breaking down the grains (mechanically and with saliva enzymes) to release the starches for easier fermentation and of adding an initial dose of microbes. As off-putting as it might sound to those of us living in the Pasteurian present, this method of starting the fermentation process in the mouth is not uncommon across the historical record—extending from at least nine thousand years ago in China, to the third century BCE in Japan, and even to traditional methods still practiced in Latin America today.

An analysis of compounds from ancient clay jars unearthed from a Neolithic village in the Henan Province of China provides some hints about these early practices. Their contents were revealed to have been a rice-based fermented beverage with hints of honey and fruit. The authors of a paper describing the discovery point out that rice does not contain the yeasts or the accessible sugars necessary to spontaneously ferment into the beverage they found traces of, concluding that saliva enzymes must have helped to start converting the foods' starch to fermentable sugars, while also introducing mouth microbes.

In Japan, a beverage known as kuchikamizake translates roughly to "mouth-chewed liquor." More than a thousand years ago, people living in Japan would chew and spit millet, buckwheat, and even acorns into the mash to ferment it into alcoholic beverages. They later incorporated rice as the starchy starter, eventually giving rise to the unmasticated sake we know today.

Some versions of a South and Central American beverage known as chicha are made in a similar fashion. The process in the Americas dates back at least to the Incas and is carried on today by the Tsimané in Bolivia and other groups. Joe Alcock, a physician at the University of New Mexico, calls the Tsimané's chicha a "frothy, delightful fermented drink." Makers chew maize or cornmeal (or another starchy substance, such as cassava) and then spit it out, mix it with water, and allow it to ferment in an earthenware pot. Some types of chicha are consumed while the fermentation is still under way—once the alcohol content is high enough to have killed off less desirable microbes but before it goes too far in that direction to do all of the microbes in. Contemporary studies of generic chicha (wild-fermented, but not necessarily a premasticated variety) revealed more than forty species of lactic acid bacteria, which were dominated by *Lactobacillus* species, including our friends *L. lactis* and *L. plantarum*, as well as *L. rossiae*—which can make vitamin B12.*

The discovery of ancient tombs on the site of the Quito international

* At least one adventurous brewer in the United States, the founder of Dogfish Head Brewery, made a batch of mouth-chewed chicha to serve at the brewery's taproom. But, he assured customers, the beer was pasteurized before the finished product was tapped. So much for the microbes.

airport in Ecuador allowed researchers to investigate the ancient microbial profile of chicha. The tombs held individuals who had been buried there around the year 680 CE—along with jewelry, clothes, food, and chicha. One microbiologist scraped sediment from the pores of clay pots that had held the beverage. From those scrapings, he was able to reanimate 1,300-year-old strains of yeasts from the ancient drink. He found no traces of the modern brewer's yeast *Saccharomyces cerevisiae*, which creates the vast majority of contemporary brews across the globe. Instead, most of the yeasts were from the *Candida* genus, including two strains from a previously unidentified type, *Candida theae*, a relative of species that live in the human mouth.†

Although these beverages might not be appearing next to the kombucha on your market's shelves any time soon, they are a fascinating—and generally safe—way of growing a human-origin microbe colony for consumption.

Wild Ales

Of course there are plenty of other ways to ferment grain besides chewing it. And we didn't have to wait until the isolation of brewer's yeast *Saccharomyces cerevisiae* to develop a wide range of fermented beverages. For generations, fermenters left their ales open to the wild microbial environment, creating a rich slurry of microbes for the consumer.

Most beers today are brewed in exceedingly sterile environments and to exacting specifications. To make beer, grains are cracked before being made into a hot watery mash, helping to release their carbohydrates.

† And yes, the researcher used these revived strains to cultivate a batch of experimental chicha, which he reports tasted just fine but left him with a terrible headache. The quantity consumed was not detailed.

Then, the liquid—now known as wort—is drained off and further heated, often with the addition of flavoring, such as with hops. After the wort has cooled, the *S. cerevisiae* yeast is added, and the mixture is left to ferment at a specified temperature and for a precise amount of time, depending on the type of beer being made.

In today's commercial processes, not just any old type of *S. cerevisiae* will do. Very specific strains are used for each type of beer. And heaven forbid any other type of microbe enter the picture. As one team of researchers put it, "for 99 percent of the beers on this planet, *Saccharomyces* is the sole microbial component, and any deviation is considered a flaw." And just to be safe, the products are usually pasteurized and/or filtered to prevent microorganisms from continuing to alter the flavor or physical characteristics of the beverage before it arrives at its destination. All of this precision ensures that your bottle of name-brand beer tastes the same each time. But it forfeits much of the complexity and liveliness of brews past.

Some brews today still contain yeasts and other microorganisms when they reach the consumer. Brewers can skip the pasteurization step if their kegs are going to be distributed along a relatively short and consistently chilled supply chain. Keeping the beer cold ensures any yeast inside will stay dormant and not continue fermenting beyond the intentions of the brewer. Other beers, known as cask- or bottle-conditioned, undergo a second round of fermentation with the addition of more yeast before they're sold.

Some brewers are using even wilder methods that harken back to beer's origins, thousands of years ago. Many brewers of lambic, sour, or wild ales aren't tied to brewer's yeast—or starter strains at all. Instead they rely on the brewery environment to inoculate their mixture and start the fermentation process. The yeasts can include *Kloeckera apiculata* (known as a spoilage yeast in most other brews) and the aptly named *Brettanomyces lambicus,* among others, as well as bacteria. This type of beer generally requires a much longer aging process, fermenting in barrels for as long as three years. During this time, various successions of microbes become prominent, ultimately favoring those that provide its characteristic sour flavor.

Beer-style beverages can be found around the world. In Korea, grains inoculated with a blend of bacteria, yeast, and other fungi are used as a starter for various products, including makgeolli. Makgeolli is a fermented alcoholic rice drink that contains live microorganisms, including yeasts and numerous lactic acid bacteria.[‡]

In Turkey and countries in the Balkan region, a fermented drink known as boza has been popular for centuries. Its basis can be millet, corn, wheat, or other grains, either fermented spontaneously or with backslop from a previous batch. Researchers sampling just three different commercially purchased boza drinks found at least a dozen species of lactic acid bacteria and eight different species of yeasts. Another team of researchers set out to plumb the murky depths of this cloudy beverage for possible probiotics. There they found strains that included known probiotic *Lactobacillus rhamnosus* and others adept at surviving in the gastrointestinal tract and helping to fend off pathogens.

Throughout Africa, numerous beer-like drinks are made and enjoyed locally. In Eastern Africa, the nonalcoholic drink or porridge togwa is made by boiling maize, millet, sorghum, or cassava flour and then fermenting before diluting it into a beverage. It contains at least a half dozen different lactic acid bacteria (including the sometimes probiotic *Lactobacillus brevis* and *Lactobacillus fermentum*) and four species of yeast and still others that add extra folate for those who consume the product.

‡ One Korean researcher I spoke with cautioned that although consuming these additional microorganisms should be beneficial, he cannot fully recommend it, as it also does contain alcohol—not, overall, the biggest plus on the health side.

In the Himalayas, a local brew known as chyang is made with fermented millet, barley, or rice. It is thought to help fight against colds and allergies. Perhaps that's one of the reasons it is said to be a favorite beverage of the mythical Yeti.

• • • • •

With all of these microbes contained in boozy brews, it might be tempting to skip the kefir and pour a pint of something more social. Alas, there hasn't been a lot of research into the net benefit of consuming microbe-filled alcoholic drinks. There has, however, been some work done on the impact of consuming alcohol on the gut microbiome. And it's not great.

Even moderate alcohol intake can lead to dysbiosis (microbial imbalance) in the large intestine, and heavy intake can spur harmful overgrowth of bacteria in the small intestine. Drinking also increases the permeability of the gut lining. And as you'll recall, having microbes leak out of our guts and into the bloodstream isn't ideal. Microbes on the loose prompts the immune system to engage, resulting in increased inflammation.

Leaky gut has also been implicated in the link between alcohol consumption and liver damage, as these escaped microbes can ultimately wind up in the liver. And as in the bloodstream, the presence of the microbes in the liver calls the immune system to attack, creating additional inflammation and, ultimately, scarring of the liver.

So, as far as your microbes are concerned, abstaining might be best, but drinking in moderation is preferable to excessive consumption. But there are plenty of other ways to boost your microbial intake—and they don't all require imbibing.

Moldy Rice

Rice has been a dietary staple in many regions around the world for thousands of years. It has also played an unsung but crucial part in creating many of the most revered fermented products. Why? It is a great place to grow mold.

In Western kitchens, food covered with mold is typically discarded without a second thought. But not everyone is so draconian about fungal spores. Rice molded with *Aspergillus oryzae,* a domesticated fungus, is responsible for countless Asian fermented foods. It—along with the grains it is grown on—is known in Japanese as koji (which means something like "bloom of mold"). Koji makes the world go round, at least in traditional Japanese cuisine. "If you didn't have any koji, the Japanese pantry wouldn't exist," proclaims food author and chef Yukari Sakamoto. It is responsible for miso, soy sauce, and sake, to name a few. Its powers lie in its ability to break down the starches in rice and soybeans so they can be better fermented.

Growing koji, although not a long process, requires substantial dedication and vigilance. This is not quite the crock-it-and-forget-it world of sauerkraut. Koji is made by inoculating steamed white rice with the mold. This often done with a shaker, dusting a layer of spores over the new rice. Some traditional makers even sing to the koji beds as they dust them. After the mixture is left in a warm space overnight, the grains are stirred and left for another night. The finished product is a collection of rice grains, each covered in delicate white, sweet-smelling mold. They resemble something like a puffed rice cereal. Eaten on their own, they taste subtly sweet and earthy.

Some of the earliest documentation of koji is from China more than two thousand years ago. The molded rice then made its way to Japan some thirteen hundred years ago and then spread throughout the region. Geneticists have studied strains of the mold *A. oryzae* from around

Asia, as well as the most closely related wild species, to try to better understand the history of this fungus that launched so many microbially rich foods. They discovered that all of today's strains have a common ancestor and thus probably stem from just one early domestication.

Humans have been working with *A. oryzae* for so long and have employed it in such a wide variety of foods that quite distinct strains have emerged to be specially adapted to each purpose. For example, sake koji is adept at breaking down the proteins in sake rice but not those in soybeans. So using a sake koji to try make miso or soy sauce won't be nearly as effective as a koji that has evolved to create those particular products.

In traditional sake, white rice was washed and steamed before the appropriate type of koji mold is introduced. Once properly started, this rice was mixed with water to create a fermenting mash. After fermenting, the contents were then poured into cloth sacks and pressed, separating the filtered sake from the cake-like saki lees (themselves used as a distinctive substrate for pickling, creating kasuzuke). Most sake today is made by more mechanized process—and also pasteurized. Some unpasteurized sake, however, can still be found, with its microbes intact.

In addition to its role in making sake, koji is also responsible for the creation of amazake, a sweet fermented rice drink. In most commercial preparations, it is boiled for pasteurization and to stop the fermentation (otherwise it will turn from a sweet beverage into an alcoholic one). But it can also be consumed before boiling, with microbes in place. It is often served at festivals, for dessert, or as baby food. It is also a popular sweetener in Japan that is seen as more traditional and healthful than refined sugars. Amazake mash can also be used as a pickling medium for produce.

ELIZABETH ANDOH'S KOJI BLACK SESAME ICE

Amazake can be purchased at specialty shops and kept in the freezer. Elizabeth Andoh, the Tokyo-based American food writer and cook, uses it beautifully in numerous dessert recipes, including this ice.

The ingredient list for this deliciously creamy black sesame sorbet, featured in her book *Washoku: Recipes from the Japanese Home Kitchen*, is short: amazake, black sesame paste, and soy sauce (bonus if you find some that has not been pasteurized). The only tools needed are a blender and a freezer-safe container with a lid.

Beat 1 cup amazake with a blender until smooth.

Add 1/4 cup black sesame paste and blend until incorporated.

Add 1/4 teaspoon soy sauce and pulse again, until the mixture resembles dark ice cream.

Pour contents into a lidded container, tapping it on the counter to release air bubbles.

Seal the container and chill in the freezer for four or more hours.

Scoop and serve. Don't expect it to last, though: It is "more addictive than fudge," Andoh cautions with a laugh.

It is also relatively easy to make amazake at home if you have koji (which can be purchased at some Asian markets) and a rice cooker. Start by making a rice porridge in the rice cooker, then add koji along with a little additional water. Allow the contents to incubate with the lid open. If you're looking to fine-tune, you can even shop around for just the right variety of koji.

Food for Microbes

In their unfermented state, many grains also offer important prebiotic food for beneficial microbes in our own guts. Wheat, barley, and other grains contain oligofructose (or fructooligosaccharides, FOS). These long-chain carbohydrates, like other prebiotics, are not broken down by our own bodies and arrive in the colon intact—and ready to be fermented by friendly local bacteria. They are known to increase the number of bifidobacteria and to boost the quantity of healthful short-chain fatty acids produced in the gut. By helping to increase the acidity of the gut, they also improve mineral absorption.

One study found that a dietary change as small as eating about one and a half ounces of whole-grain breakfast cereal for three weeks prompted the growth of much larger populations of bifidobacteria and lactobacilli in the gut. Now, that's not such a difficult shift to make.

Whole grains, including oats, barley, rice, and corn, are also a solid source of prebiotic resistant starches. These are the starches that resist digestion by our body's own enzymes (unlike many other forms of simple starches, which can be rapidly broken down into simple sugars). The handy microbes in our guts have evolved the enzymes to break down these fibers, releasing food for themselves—and through that, also producing beneficial compounds for us.

In addition to this naturally occurring form of resistant starch, there is also retrograded resistant starch. This compound comes from otherwise simple-starch foods (pasta, rice, potatoes) that have been cooked and then cooled. The crystallized starches are too tough for our digestive system, leaving the carbohydrate chains intact for arrival in the large intestine.

So some otherwise overlooked sides, such as pasta salad, potato salad, and other cooked and chilled starchy dishes, might be better for us and our microbes than we had imagined.

Yet another food that might fit this bill is sushi rice. When the rice has been cooled, these retrograded starches form, stocking food for your microbes. In a proper sushi restaurant, sushi rice is often served closer to room temperature, and if it hasn't been fully chilled first, it might contain fewer of these helpful compounds. But the research into specific temperatures required for retrograding starch in various foods itself has not yet crystallized. So when it comes to feeding microbes, it seems that rice, pasta, and potatoes are best served cold.

• • • • •

You can take your pick of grain approaches: prebiotic whole grains, a side of picnic potato salad, or some chicha. But if the orally inoculated beverages are giving you pause, you might want to steel yourself—and take a quick swig of the nearest non-masticated beverage—before proceeding to our next fermentation destination.

We have been living in the world of the acid, the sour ferment: lambic beers, sauerkraut, yogurt. But there is a whole category of foods that take fermentation in a very different direction: gooey, alkaline rotting beans.

CHAPTER SEVEN

• • • • •

Basic Beans
Legumes and Seeds

Many foods, like dairy and produce, are imminently perishable, so it makes sense that various cultures have developed ways to control the spoilage process so they can be consumed over a longer period of time. Legumes and seeds, on the other hand, can be readily dried and stored, making it somewhat curious that they have also long been objects of fermentation. To boot, they create some of the more challenging fermented foods.

Perhaps chalk it up to the persistent power of human curiosity, creativity, and forgetfulness.

Breakfast of Champions

Japan is renowned for its healthy, long-lived populations. The southern Japanese island of Okinawa, for example, boasts more people per capita over the age of one hundred than just about anywhere else on Earth. And many of these centenarians still live independently at home. Explanations for this long and healthy life range from a tradition of honoring elders; to a balanced lifestyle; to a longevity-boosting diet, rich in fish, plant products, and fermented foods. In Japanese culture, beyond pickles and koji, fermentation has also been employed to transform a key staple: soy.

In Japan and throughout East Asia, the soybean has long been a crucial element of diet, nutrition, and cuisine. It is consumed in the form of edamame, tofu, miso, and soy sauce. And then there is natto.

• • • • •

I was surely not the first Westerner to be warned about natto. A pungent, gooey dish, natto resembles slimy, rotten beans because, well, that's basically what it is. It is a goopy topping, flavored, some suggest, like a very rank cheese. To tens of millions of people, it is also part of a nutritious breakfast.

And this is exactly how I first encountered it, jet lagged from a twenty-hour trip, on my first morning in Tokyo. To my excitement and mild trepidation I found a small, commercially packaged cup of natto on my hotel breakfast tray. Approaching it with a certain amount of caution, I opened the lid and prepared for an olfactory punch.

But none came. I gave it a quick stir with my chopsticks and attempted to scrape it out onto my small bowl of rice. Vexing strands of ooze refused to relinquish their hold, stretching stubbornly the length of my reach. I looked around the small dining area for guidance. No one else seemed to be battling the filaments. The breakfast was already presenting challenges—especially as a nonnative user of chopsticks and equipped only with tiny waxy paper sheets for napkins—and I hadn't even brought a bite to my mouth.

So I was relieved—perhaps even a bit disappointed—when I finally managed that first bite to find the natto neither pungent nor revolting.

In fact, on second consideration, I found it actually quite tasty, carrying an umami flavor and faint hint of sweetness.

But perhaps I had missed out on the full experience that first meal. That afternoon, a local interpreter explained to me that the proper first step in eating natto is to stir it with the tips of your chopsticks—a lot. This activates more of the gooeyness and more of its flavor. In fact, according to famous Japanese food aficionado and potter Kitaoji Rosanjin, one must stir natto 424 times* to release the very best in this food.

But before the 424 turns, natto has a fascinating journey, including a fermentation that is at once surprising and basic.

• • • • •

Natto begins with whole soybeans, which are washed and soaked. They are then steamed before being left to ferment at warm temperatures for just one day. The brief process doesn't sound like it could possibly be all that transformational—and certainly doesn't preserve the harvest as much as other fermentation processes or even drying would. So there must be a reason this food has taken hold—and shown such lasting power.

Those who partake are rewarded with a healthy serving of *Bacillus subtilis*, a bacterium that helps stimulate the immune system and can aid in battling gastrointestinal and urinary tract illnesses.

It is this bacterium that is behind much of natto's funk. You see, *B. subtilis* is a bacterium of a very different sort than our lactic-acid-producing friends of the previous chapters. It is an engine of alkaline

* There is even a Japanese gizmo that will do this for you—adding soy sauce, if you desire, at just the right time: stir 305. I didn't invest in the gadget, but I did give this prescription a whirl back home with a cup of natto from a specialty market. After about 100 turns, the natto developed a much thicker mucus between the beans. And a lingering acrid aftertaste. Perhaps the true connoisseurs can tell the difference 324 stirs later.

fermentation. While we were diving into the acid ferments—from kefir to sauerkraut—an entirely different side of the fermentation spectrum has been quietly, flavorfully, waiting in the wings.

But don't fear for the safety of this alkaline-fueling microbe in our acidic digestive tract. Its work might create the basic environment of natto, but it is perfectly capable of surviving ingestion and digestion. *B. subtilis*, it turns out, is anything but subtle—in flavor compounds or in lifestyle. Unlike sensitive *Lactobacillus* species, *B. subtilis* is renowned for its ability to stay viable in extreme environments, including our acidic stomachs and high-temperature baking. It can even survive for up to six years in space.

• • • • •

So how did this intrepid bacterium come to unite with innocuous steamed soybeans—in the process, transforming them into something so different?

The Japanese have been feasting on natto for centuries. Legend describes its accidental discovery (perhaps in 1083 during the Gosannen War), when warriors left their boiled soybeans sitting too long in rice straw.† Apparently the warriors found the results so appealing that they shared some of the beans with a famous general. And so the process carried on, stuffing boiled beans into straw, for hundreds of years. In the early 1900s, with the advent of microbiology, producers found that they could create a starter culture out of isolated *B. subtilis*, eliminating the need for actual barnyard straw in the fermenting process.

Today a lot of natto is made under quite controlled conditions, but one factory still does much of the work by hand.

---

† *B. subtilis*, sometimes known as the hay bacillus, is a common resident of soil (and thus hays and grasses).

The Tengu Natto factory is in Mito, the capital city of Ibaraki Prefecture and the heart of the natto world. The factory is just a short walk from the central train station on an unremarkable road next door to a tire store. The small building still smells of damp straw, and on a quiet afternoon, no shopkeepers are to be found in the small front shop. Wandering beyond the small storefront and down a hallway, you can see the glassed-off production space. There, workers still wrap their beans for fermenting in carefully chosen straw and pack crates by hand. Upstairs is a one-room natto museum, detailing the food's history. Natto made at Tengu hasn't changed much over the company's century of making these distinctive beans. Before I leave, I find a Tengu worker and buy a traditional straw-wrapped parcel of natto to take back with me. And wait to see what mysteries it will reveal.

• • • • •

Japan's Ibaraki Prefecture might be the epicenter of natto, but the dish is popular throughout the country. And to find a mind-boggling array of natto, those in Tokyo can visit the Ibaraki antenna shop. There are dozens of types—from small beans to big, and from traditional straw packing to convenient cardboard containers. And there are nearly as many other natto-based products, including natto sauce, natto kimchi, and prechopped natto ready to put into sushi hand rolls. There are even dried natto beans, a snack food known as hoshi natto, which regains some of its gooiness as you chew it.

• • • • •

Finding natto to be such an intriguing fermented dish, I began collecting different types during my stay in Japan, stashing them in the mini-fridge in my traditional *ryokan* room. By the time the last night of my stay came around, I had accumulated numerous samples—from plastic packages picked up on a train station platform to traditional straw-wrapped bundles straight from the makers. During my time in Japan, I marveled at the number of dishes that accompanied every meal—small dishes offering a wide array of flavors, nutrients, and experiences. Knowing that this variety is a mainstay in most Japanese meals, I nevertheless decided to commit Japanese culinary sacrilege and treat myself to an all-natto dinner.

To start off my unconventional meal, I open a small round container from the basement of a department store in Mito, which comes with soy sauce and spicy mustard. It is quite goopy but mild and familiar. The addition of the soy sauce makes it almost sweet. Next is a small straw-wrapped natto from an antenna shop. This one was sold in a plastic package with holes in it for air exchange. Before tucking into the beans, I can feel and see the natto inside through the straw wrapping. Wherever I decide to pull apart the strands of straw, the natto comes poking out. Unsure of the protocol (and eating in the privacy of my own room), I decide to eat it right out of the straw bundle, digging out each bite with chopsticks. It is not quite as mild as the first, but it is quite tasty. The following course is a large straw-wrapped natto package purchased from the Tengu Natto shop in Mito. This one carries quite a mild flavor, although getting through all of the stiff straw layers is a challenge. Finally, a rectangular box of natto mixed with what seem to be chopped vegetables purchased on a train station platform is the only natto that I cannot conquer. It is, alas, too pungent.

Picking through an unconventional dinner of this curious food only deepens its allure. Even just a few bites make for an immersive, dynamic eating experience. The beans themselves are partially deflated and have variable textures. And the beans are truly interesting, occasionally moving in front of you long after you've put down your chopsticks. They seem almost animated. As Tengu Natto proclaims on its website, "Natto is alive!" It appears to be true.

JAPANESE NATTO AND RICE

Most Japanese meals consist of many small dishes, but a serving of natto and rice can take center stage for a filling breakfast. Additions of chopped green onions, nori, and sometimes even a raw egg make for a fiber-, protein-, and vitamin-and-mineral-filled meal. For this simple version, you will need natto (available at many Asian groceries, sometimes in the freezer section), soy sauce, chopped green onions (a little prebiotic boost), nori, and cooked white rice.

Place about a third of a cup of natto in a small bowl.

Stir with chopsticks for at least 30 seconds; 305 rotations is the precise number of stirs recommended for this step, which takes roughly two and a half minutes if you're feeling patient (but not up for counting).

Add about half a teaspoon of soy sauce.

Stir some more—at least until incorporated, or as many as 119 turns (about a minute) if you are feeling ambitious.

Pour the mixture over a small bowl of white rice.

Add a sprinkling of chopped green onions and crumbled nori, and enjoy.

You can also make your own natto with a natto starter purchased from a specialty store—or with natto that still contains its live microbes. Soak dried soybeans overnight in ample water. Boil the beans until they are soft; this can take several hours, so many people instead steam them in a pressure cooker to cut the cooking time to about 45 minutes. Spread the beans on a rimmed baking sheet and sprinkle with the starter. Cover the beans (straw wrapping optional) and set them aside to ferment at 85 to 105 degrees Fahrenheit for roughly 24 hours (or longer for a stickier, smellier product). Voilà, your own stringy, smelly beans.

Basic Beans—and Seeds

Japan is not the only country to create such an improbable food. Korea has a similar food called cheonggukjang, prepared after the soybean harvests in the fall and winter. Records of its creation date back to the seventh century CE In the Himalayas, *tungrymbai* is a regional version of natto. Tungrymbai is a particularly sticky food, made by fermenting cooked soybeans in a bamboo basket that has been lined with leaves of a local plant for a few days. The result has been described as "a brown mass with a characteristic odor," according to one diplomatic researcher. But in this slimy alkaline sludge there lurk many different microbes, including *B. subtilis*, as well as *Enterococcus faecium* (some strains of which are probiotic), and various yeasts, including *Geotrichum candidum* (also responsible for the distinctive texture of Finland's yogurt-like viili).

But whole fermented beans need not be as sticky or pungent as cheonggukjang or natto. In China, a fermented black soybean known as *douchi* provides a popular flavoring. These beans retain more of their structural integrity and are far less gooey. And they have clearly long been a local favorite: douchi was discovered buried in an ancient Han dynasty Chinese tomb from the first century BCE.

In many places in Africa, alkaline ferments of legumes and seeds appear throughout different cuisines. Perhaps the ingredient most commonly used for fermentation in West Africa is the locust bean, which is a good source of protein and vitamins, and despite its name, is actually a seed (which resembles an irregular lentil). These fermented seeds become *sumbala,* which is popular throughout the region.

The process of making sumbala is labor intensive, but the results are highly nutritious. Women have traditionally harvested the seedpods. Those pods are then cracked open, and the seeds are pounded to separate the hull from the insides. The hulled seeds are then left to dry before starting the fermentation process (or stored for later batches). Dried seeds are cooked and then pounded again and dried. They are washed and then boiled yet again. After this final cooking, they are poured into fabric bags. Weights are placed on the bags and the contents are left to ferment for about three days. The final pasty product is then made into small balls for storage.

Researchers point out that at the outset of the process, the substantial cooking of the seeds likely reduces the amount of microorganisms present on them to start the ferment. But they note that microbes are likely introduced through the processing implements, the makers' hands—and even the air. Like natto, sumbala goes through an alkaline fermentation, the result being less stringy but still somewhat gooey. Microbial analysis has found that this strongly flavored food may contain various *Bacillus* species as well as numerous fungi (including *Penicillium* and various *Aspergillus* species). It is eaten frequently as a condiment and to lend flavor to stews and other dishes. Although this flavorful, nutritious food, still made using ancient processes, may be waning in popularity, writes a team of researchers: "They are associated with some problems, such as having a short shelf life, objectionable packaging material, and characteristic putrid odor and stickiness." Perhaps further discoveries about their microbial benefits will help fuel additional sustained appeal.

Mold

Clean, pressed, plastic-sealed blocks of tempeh available at the grocery store, neatly elide this fermented soy's elaborate, yeasty origins. This popular protein-filled food is actually held together by a matrix of spores—even in that tame package you have in the fridge.

In Indonesia, where this food has been eaten for generations, locals enjoy a variety of these fermented bean patties. To make tempeh, soybeans are soaked and cooked, and their hulls are removed. Originally, the beans were likely pressed in hibiscus leaves, which naturally contain spores of *Rhizopus oligosporus*,‡ and then left to ferment. (It is now more common to add it or *Rhizopus oryzae* directly in a more controlled environment.) Then the mixture is spread out and cultured for a day or so in warm ambient temperatures. During this time, the fungus grows between the beans, adhering them with its whitish mycelium. The product is then packed up and kept for later consumption. Various other microbes also lurk in fermented tempeh, including *Klebsiella pneumoniae*, *L. plantarum*, *Enterococcus faecium* (which has been shown to inhibit the growth of listeria), and other bacteria and yeasts, making the traditional stuff a lively, microbe-filled food.

• • • • •

Within all of these alkaline and moldy ferments, there is a terrific world of microbes to be discovered and consumed. But unlike dairy and many forms of produce, beans are of course, quite easily dried for long-term storage. Which begs the question: Why go through the trouble of fermenting them—with all of these fuzzy yeasts and funky bacteria?

In addition to adding microbes to the diet, these cultured beans also wind up with a different nutrient profile than the raw or cooked versions. In the fermenting process, the microbes help to break down fibers and proteins and may even synthesize additional vitamins. In lab simulations, soy-based tempeh even promoted the proliferation of *Bifidobacterium* bacteria in the gut. Although none of this was explicitly apparent

---

‡ Which has been linked to reduced GI infections, lower cholesterol levels, slowed tumor growth, and other health benefits. The fungus has even been shown to be able to treat wastewater.

to our ancestors, it might have made itself known in additional, subtle health benefits.

Fermented soy doesn't end, of course with tempeh and natto. It also lives many other lives.

Ancient Barrels

Miso, a paste made of ground fermented soybeans, is thought to have originated in China—possibly some 5,000 years ago. And it has been a mainstay in Japanese diets for at least the past 1,500 years, arriving on the island roughly the same time as Buddhism. In its simplest serving, a modest amount of miso is spooned into heated water and mixed, perhaps with a pinch of chopped scallions (a prebiotic touch) and a few small cubes of tofu—not unlike the ubiquitous soup at Japanese restaurants around the world.

In Japan, however, miso is not just reserved for soups. It is also used in salad dressings, as a marinade, and to add salt and umami flavor to sauces. It makes appearances in sweets, such as a Yamagata Prefecture specialty of miso with walnuts wrapped in shiso leaves. It can even be used as a medium for pickling vegetables. Miso can be made from any variety or combination of soybeans, hemp seeds, millet, rye, barley, rice, or wheat.

• • • • •

Today in Japan, you can find a dizzying array of miso varieties. In one Tokyo antenna shop, an entire refrigerated wall is packed with miso, and shoppers can sample dozens of different kinds. A barley miso is light and sweet, with slivers of intact grains. A rice miso verges on cloying. And an aged soy-only miso is thick, rich, and intense.

To make miso, the base legume or grain is fermented using the domesticated koji fungus *Aspergillus oryzae* (though many other microbes can be found there as well). The soybeans themselves contribute yeast and bacteria that prod lactic acid bacteria along in transforming the ferment. On its way to maturity, the miso paste also becomes a concentrated source of *Lactobacillus acidophilus*, which has beneficial effects on the immune system and can help ward off infections. One study examined the effects of consuming miso soup mixed with natto for two weeks and found an increase in participants' lactobacilli and bifidobacteria—as well as a boost in beneficial short-chain fatty acids.

• • • • •

In Japan, many of the miso makers use not only ancient traditions but also ancient equipment. My journey to find miso being slow-aged in hundred-year-old barrels began, improbably, in a bullet train from Tokyo Station, racing down the coastline through a blur of densely populated high-rise cities and suburbs. After various commuter rail transfers on impeccably timed trains, I at last arrived in Okazaki. This small city in central Japan is home to one of the foremost miso makers, Hatchō Miso (Eighth Street Miso), whose products are eaten by the emperor and the public alike. The company makes its famed miso in picturesque, if unassuming, low-slung tile-roofed buildings on Eighth Street, amidst light industrial businesses. I soon found out that the magic was hiding inside.

The company has been making miso on the same street in Okazaki for some seven hundred years. Five hundred years ago, it became the favorite miso of the local prince, Tokugawa Ieyasu (whose family's sixteenth-century castle still stands just eight blocks away).[§] Tokugawa

§ Tokugawa's life was also the basis for the popular 1975 novel *Shōgun* and subsequent adaptations.

would go on to become a famous shogun, feeding his troops, so the legend goes, Hatchō miso as a key component of their rations. Once he rose to power and moved his government to Tokyo, he insisted his childhood miso be sent from the same Okazaki factory, earning the company the status of official supplier to the shogun. In the late nineteenth century, Hatchō also became the official miso of the emperor of Japan and remains so to this day. Hatchō is now run by Kyuemon Hayakawa, who is said to be the eighteenth generation of his family to oversee the company.

To begin the miso-making process, soybeans are washed, soaked, and steamed for about two hours. They are then crushed and inoculated with koji spores—and not just any koji spores. These are certainly not interchangeable with koji used to make soy sauce or even with koji from other miso makers. Cultivated for centuries in-house, Hatchō's koji spores are said to have evolved into a strain unique to the company, its processes, and its environment. These special koji spores incubate with the beans for about three days. Sea salt and a little bit of water are added to the mixture, which is then packed tightly into wooden barrels. (Formerly, workers stomped on the paste with their feet to get the air out.) The factory ages its miso in massive barrels that are used continuously for about a century before being retired. The cedar barrels, held together solely by braided bamboo, reach seven feet tall. Once filled, the barrels are covered with wooden lids piled with pyramids of large, smooth river stones, then left to ferment.[¶]

The barrel contents age for twenty-four to thirty months in single-story wooden barnlike warehouses. The hulking stone-topped barrels are perched on short platforms and arranged in rows, resembling a

¶ The stone piles themselves are legendary. Thousands of pounds of the carefully selected rocks are stacked carefully on top of each large miso barrel to provide pressure for proper fermentation. They say that these piles are so artfully constructed that they never fall—even during earthquakes.

labyrinth of outsize upright wine casks. There was a sweet smell to the dark aging buildings, which had large doors flung open to the overcast April day.

Hatchō's basic process for making miso is no different from that of most other makers. But the creation of this particular miso, with its exceptionally rich flavor and thick consistency, may well be linked—microbially—to these historic wooden barrels and old wooden buildings, embedded in the physical environment, literally lurking in the rafters.

• • • • •

Miso is so tied to its environment that *temae miso*—"my own miso"—is a common phrase in Japanese. It literally means the miso that you have made with your own hands in your own home—because there is in fact such a thing an individual person's miso. "You used the exact same ingredients—the exact same beans and the exact same salt and everything—but still all the tastes would differ because of that different bacteria that's floating in the homes, as well as on your body," Tokyo chef Nobuaki Fushiki explains to me through an interpreter. "That's why everybody likes their own miso. So in everyday conversation, when you say something's *temae miso,* it means, 'I'm commending myself.'" A microbial basis for giving yourself a little pat on the back.

Indeed, some chefs are thinking deeply about what it means to make their own miso. In New York City, David Chang of Momofuku and his colleagues have been experimenting with miso in their test kitchen, and he knows that it is very much *their* miso. "Even if we all followed precisely the same recipes and used the same products, our products would differ, and reflect the nuances and complexity of our microbial terroir," he and his colleagues write in a paper published in the *International Journal of Gastronomy and Food Science.* "We are on the cusp of a movement that connects us not only to these ancient techniques but to our environment on the deepest possible level."

Not-So-Simple Sauce

Another bean brew of sorts is hiding in plain sight: soy sauce.

Soy sauce might seem to be a relatively simple condiment.** I always figured it to be some sort of mix of salt, soy extract, and perhaps colors and sweeteners. And in many commercial blends, such as the ones you pick up in the supermarket, that's not too far off. But in its traditional preparation, soy sauce is one of the more complex legume ferments, relying on a succession of mold, bacteria, and yeast to produce its rich and subtle flavors. And served in its original form, it is chock-full of varied microbes.

• • • • •

To make Japanese soy sauce, known as shoyu, the process starts with our fungal friend koji. For traditional soy sauce, the well-tuned koji is mixed in with whole cooked beans and roasted grains. The mold is then allowed to spread and mature on the beans for two to three days before the fuzzy landscape is blended with salted water.

A subsequent fermentation process occurs in stages. First, the koji and a handful of other microbes dominate the fermentation process. Then other microbes from the environment arrive, which can include

** Soy sauce, like so many of these foods, is hardly a single product. The varieties among—and within—countries are seemingly endless, with more than a dozen basic types in Japan alone. In Japan, tamari is made with a base of soybeans and has little if any wheat. Shiro, on the other hand, is mainly wheat-based. The most common soy sauce in Japan falls somewhere in the middle. Known as koikuchi, its base is roughly half soy and half wheat.

Staphylococcus, Lactobacillus, and *Bacillus* species. Finally yeasts such as *Zygosaccharomyces rouxii* (which is responsible for some of the signature flavors of soy sauce) hold forth. A graph of these overlapping surges and falls in populations over time looks like a cross section of ancient geological strata. This is no two-strain party.

This old-style way of making soy sauce actually creates two products. Once the fermentation process is complete, the mash, known as moromi, is squeezed through cloth or filters. The liquid that runs out is the soy sauce, and the solids that remain from the moromi can be used as a rich spread akin to miso, or as flavoring for sauces. In fact, soy sauce may well have had its origins in miso production, the soy sauce being the liquid that accumulated during production. Soy sauce purists will undoubtedly argue it was the other way around.

Like miso and so many other long-standing ferments, soy sauce has important ties to place and history. After the 2011 earthquake off the coast of Japan, the resulting tsunami wiped out the Yagisawa Shoten soy sauce plant, which had belonged to—and been run by—the same family since 1807. With their entire inventory and production area destroyed, they lost not only their product and infrastructure, but also their specific environment and cultures, the strains of organisms that had made their soy sauce *theirs.* But later a single remaining bottle was discovered. It had previously been donated by the company to a local medical university for research and had somehow survived even the university lab's destruction. Upon its unearthing, the cultures inside were returned to the owner of Yagisawa Shoten. And production of the old sauce started once again.

Fermentation Culture

In Japan, fermentation is not just about making miso, natto, and nukazuke. Fermentation is, as chef Nobuaki Fushiki exemplifies, something of a cultural obsession. Fushiki is the chef and creator of Shiojiri Jozojo, a fermentation-based restaurant in Tokyo, where he melds ancient techniques of microbial alteration with new culinary approaches of his own design. The name of his restaurant references the word for brewery, which he applies to describe his many processes—and which extend far beyond the Japanese sake and beer on offer.

• • • • •

His restaurant perches on a swoop of a sleek, buttoned-up street in a quiet portion of Tokyo's Shibuya ward. The interior space looks at first glance like something that would pass for a small neighborhood tapas restaurant. But right away, visitors are tipped off that this place is different. A counter just inside the door is lined with jars of fermenting soy sauce, teas, vegetables, and fruit. A string hung along a beam holds drying morsels of some kind. An air of mystery fills the cozy and modest dining room.

On the night I visit, straight off of a long-haul flight, the restaurant is officially closed. But Fushiki has prepared an elaborate dinner for me and a couple of local acquaintances—who also help out as informal translators and commentators.

For starters, we receive a miso soup made with unpasteurized miso and grated potato, which has a subtle umami, meaty flavor. Next is Fushiki's take on sushi, in which rice covers a slice of tuna and an avocado miso paste that has been fermenting for four days. A caramelized fermented soy sauce brightens a layered pudding-like soup that reveals egg curd and a layer of fermented river eel guts beneath a sprinkled bonito flake and spicy pepper topping. Following that dish is a ball of fermented black rice with balsamic vinegar, wrapped in fermented yellowtail tuna, and topped with a thin radish slice, scallions, and fish roe; it carries a delicately complex flavor, with the fermented rice lending a slightly turned-but-not-bad taste. The final item of the course is a tuna slice topped with tomatoes, aged with garlic and olive oil.

That, apparently, is just the beginning. The next course arrives on a lovely lacquered wooden platter, with thirteen different items, from pickled beans to marinated mackerel to aged tofu to fermented plums to liver fermented with natto. Fushiki's dishes are each artfully assembled and balanced in flavor—even if some of those flavors, such as those from the fermented squid guts, are rather strong. He has an expansive and playful mind, taking inspiration from old Japanese traditions and contemporary global culture. As we discuss the dishes, he says listening to Michael Jackson's "Thriller" recently got him thinking that he was making a kind of "zombie food," reanimating it with live bacteria.

A second large tray comes, displaying the most beautiful culinary presentation I have ever seen. The landscape is like a springtime diorama, sprinkled with tiny flower petals and anchored by a miniature treelike arrangement of Japanese basil branches in a small cup—brought down to earth by a "drunk shrimp" (aged in sake) peering out of it. Of the nine different items, fish abound, but vegetable creations do not take a back seat. A dried fermented radish is soft and meaty. A different radish has been fermented with cherry blossom leaves, giving it an amazingly fresh flavor, its tones of bitterness balanced out by slight sweet notes of flowers and honey. A pile of aged seaweed offers a true taste of the ocean. As someone who appreciates the ease of preparing one-pot dinners and who once earned the affectionate nickname of "lazy chef," my mind boggles. When I remark that he must be unspeakably busy preparing all of these incredibly intricate morsels (which he delivers to the table himself), Fushiki humbly jokes, "I have so many cooks—the bacteria!"

And his bacterial bacchanal is not yet finished. An interim soup of minced shrimp in broth with fiddlehead marinated in yeast is a heavenly, light, and fluffy pause—before a chicken skewer marinated in miso, thyme, and red wine emerge, served with grated horseradish and lily buds.

Finally, and at last, for a simple and unpretentious dessert, he brings a small strawberry drink. This is amazake, the traditional Japanese beverage made from koji-fermented rice. Fushiki's version is subtly sweet from the fermented rice, balanced with a slight chalkiness of yogurt, rounded out with the fresh summer flavor of strawberry.

• • • • •

Fushiki fancies himself a brewer—a term that in Japanese extends beyond the making of beer. He sees "cooking" with microbes as a way to dip into the country's deep culinary traditions while also innovating with flavors, textures, and presentation. Several of his dishes, such as the miso-marinated chicken skewer, do get cooked, destroying at least some of the microbial life. But he estimates that a good two-thirds of his fermented dishes are served raw, their organisms still intact. This is, however, not his top priority. "What I value very much is the umami and the sweetness and the flavors that microbes make—that a human cannot reproduce and create," he says. As a result, the cuisine might turn out to be healthful. But flavor, not health, is his first priority. As he jokes, "Take, for example, macrobiotic raw food—you have health as the priority, so it doesn't taste good," he says, following up with his abundant joyful laughter.

Allowing so much of his creation to be worked on by microbes also introduces some element of unpredictability. So I ask him if it is challenging to work with such a variable process. "As a Japanese, you're sort of accustomed to this variation caused by the different fermentation process," he says. "Enjoying the variation is part of enjoying the fermented food." When I ask if he sees his approach to cooking as more of an art or a science, he says he calls his work a blend of science and aesthetics.

So as a scientist and an aesthetician, Fushiki is a tireless experimenter.

He uses a standard natto starter for his house-made soy natto beans. Theoretically, though, he says, "this would ferment anything." So he tested it on—by his account—two hundred different items. In the end, he says, the best result was still using cooked soybeans as fodder for those specially trained microbes.

And there is a good reason for that. Any organism that has been used again and again for the same purpose begins to adapt to that process—whether plant, animal, fungus, or bacterium. And although we don't think of microbes as we do corn or cows, we have domesticated many strains, bringing them from their wild states—living on plants, soil, or the body—into the service of helpful digestion of our chosen foods. Natto culture is just one example. Koji is another. In the midst of explaining this, Fushiki dashes into the back room and reemerges with a large glass jar containing a lumpy, thick, earthy brown paste. It is his own six-year miso, which he insists we try. It is impossibly dark and rich. An entire microbe-made universe away from the stuff in a plastic tub sitting in my refrigerator at home.

Fushiki also has a thing for soy sauce. Really old soy sauce. It is not easy to find traditionally made unpasteurized sauce, he says, so he makes his own. "*This* is soy sauce," he says, taking out his three-year, (literally) handmade sauce, which is fermented in a large ceramic vessel. He mixes a fermented soybean base with salted water and lets the product age. "Mix it every two weeks," he instructs me. "Use your hand to do it so that it would have lots of your own bacteria that will seep into the soy sauce." It is rich and sweet.

He follows this with his *six*-year soy sauce, kept in a large glass jug. This is not soy sauce as I have ever known it. It seems like a totally different food. It is thick and richly umami. The six-year sauce requires extra attention, he says. Mid-fermentation, the mixture is squeezed, draining the liquid off, at which point he adds more yeast. Over time, it becomes very thick, almost like miso.

One of the hallmarks of many of these fermented soybean products is the browning process usually associated with cooking, known as the Maillard reaction. In miso, this occurs when amino acids (pieces of protein) react with sugars, which have been created in the fermentation process. This explains the difference between typical brown miso and the lighter versions made with beans lacking much of their original protein.[††]

Fushiki calls himself "a Maillard reaction maniac—I react violently to anything browning!" he says laughing. As if to prove it, he says he also has (somehow) 51-year-old miso (which, for the record, is older than he is). It "smells like a horse hovel," he has said.

• NOBUAKI FUSHIKI'S MISO AND AVOCADO SPREAD •

This simple miso blend is tasty as a spread or a dip. At his restaurant, chef Nobuaki Fushiki tucks a dab into a piece of tuna sushi. The ingredients are miso, avocado, garlic, olive oil, sesame seeds. The formula below is heavy on the miso (and thus quite salty) and will keep for weeks in the fridge. For a fresher-tasting version, reverse the quantities of miso and avocado and serve immediately.

Combine with an electric mixer:

7 OUNCES OF MISO
FLESH OF ONE AVOCADO
1 GARLIC CLOVE
1/4 CUP OLIVE OIL
2 TABLESPOONS SESAME SEEDS

Blend until smooth and refrigerate until ready to serve.

†† White miso is made by vigorously boiling soybeans, removing more of the protein prior to fermentation, using additional rice or rice koji, and fermenting the result for a shorter period of time.

The fascination with fermentation runs not just through cuisine and the minds of chefs in Japan, but also throughout daily life there.

For example, there is an entire anime series—in books and on TV—devoted to the mysteries of fermentation, casting different microbes as sidekicks to the human protagonists. In the award-winning *Moyasimon: Tales of Agriculture*, Takayasu Sawaki, the teenage protagonist, has the ability to see and communicate with microorganisms. These side characters open the door to unseen worlds, uncovering a bootlegging sake scheme and locating buried kiviak (fermented seal stuffed with birds—a specialty from Greenland). The microorganisms are drawn as endearing individual characters, their names and attributes noted in the margins.[‡‡] Sawaki's constant companion is *Aspergillus oryzae*, a helpful microbe that hangs out on his shoulder.

In Japan, charming characters pop up just about everywhere—on mobile phones, on bags, practically out of the ether. There are countless cute animals, but there is also a cranky egg and a depressed burned roll. And the character *de résistance*: Nebaaru-kun,[§§] or Little Sticky Boy. He is the natto character.

Nebaaru-kun, like many characters, doesn't just exist in one frozen, smiling iteration. Nebaaru-kun pops up in myriad dynamic embodiments. Which makes sense—he is meant to represent the dynamic nature of natto. Some iterations—on key chains and plush toys—even show him with his mouth open and tongue sticking out so that it almost looks like he's signaling a cute version of disgust.

But that is unlikely the intent.

---

‡‡ Also of note is their default call to action for either food or beverage is "Let's brew!," which reflects a nuanced cultural difference in thinking about the process of fermentation and its mechanisms. This view is also reflected in Fushiki's use of the word brewery to describe his fermentation restaurant.

§§ The word *neba* in Japanese is used to describe something that is slimy.

Even without the scientific knowledge we have today about microbes—in our food and in our bodies—in Japan, fermented foods have long been regarded as healthful, which is more than enough to get over an initial unexpected taste or consistency.

• • • • •

Over lunch with author Yukari Sakamoto, she explains the simple but revolutionary approach to food there: "In Japan, you say, 'Eat it, it's good for you.'" And that's it. An appealing or gratifying taste is a secondary consideration. "The Japanese love health benefits," she says, citing a recent television show that was devoted entirely to the healthful properties of natto. It's true. When I asked about a fermented sauce in an upscale food shop, the saleswoman says, not "it's delicious," or "it tastes really good as a salad dressing." Even though both of these things are surely true, she says instead: "It's a no-brainer. It's probiotic."

This sort of thinking starts early. Children in Japan are exposed to a broad range of foods—including natto—from a young age. Goldfish cracker culture, it is not. In fact, Sakamoto says her son's first phrase was the sign for *again,* followed by the nickname for natto.[¶¶]

"I never made him special food," Sakamoto says. "He ate what we ate." And when they're away from home at school, children aren't allowed to be picky eaters, either. Teachers insist students eat whatever their parents packed for them in their bento box—and all of it, she explains.

¶¶ In the States, on the other hand, our neighbors' daughter's first expression was the sign for more and the word bacon.

Cocoa's Secret Compounds

Legumes and seeds are of course much more frequently eaten unfermented. And in this more common state they can also have a positive impact on the gut microbiome by providing important prebiotic compounds. Much of this impact comes from resistant starch, which is naturally occurring in varying amounts in lentils, kidney beans, and black-eyed peas, among other plant products. This compound is tied into the structure of the plant, which means that the more a plant is milled, cooked, or otherwise processed, the less intact it will remain for us. Or more accurately for our microbes.

Prebiotic, microbe-feeding compounds don't come only from the foods we feel like we ought to eat, though. There is also a magical bean that is good for the gut microbes: the cocoa bean.

About an hour's walk from the Swiss village of Gruyères is the town of Broc. There resides one of Switzerland's most famous chocolate factories: la Maison Cailler. Although the chocolate was sold only in Switzerland until recently, Cailler is one of the country's best-known brands, having been around since 1819. The company now uses millions of pounds of cocoa each year to make its popular confections. Their mansion-esque factory was built around the turn of the twentieth century and still stands as the company's flagship site.

Chocolate arrived in Europe only in the sixteenth century, centuries after the first wheel of Gruyère was rolled out of a cave in the Middle Ages. But the cocoa bean, native to South and Central America, had long provided a popular beverage and medicinal tonic. Genetic research has found the plant was first domesticated in what is now Peru. And residues on ancient ceramic containers suggest that humans were already consuming cocoa at least four thousand years ago. It was valued as a ceremonial drink, and its beans were even, at times, used as currency.

After cocoa's arrival in Europe, the ingredient spread, first as a frothy beverage. It took a Swiss family, however, to turn it into the chocolate bars we known today. François-Louis Cailler founded an early cocoa-processing facility in the early nineteenth century on the north shore of Lake Geneva, not far from Lausanne. But at the time, as Cailler guides will tell you, their version of chocolate was a "crumbly mess." Cailler's son-in-law Daniel Peter, a former candlemaker living nearby, had some ideas about improving it. He looked down the road to his neighbor Henri Nestlé, whose factory was making a powdered milk-based baby food. Working together, the two of them developed modern-day milk chocolate. Cailler, which was purchased by Nestlé in the 1920s, still sources its dairy from local Gruyères cows but must reach farther afield—to Central America and West Africa—for its cocoa.***

For decades, cocoa and dark chocolate consumption has been linked to numerous positive health effects, such as increased insulin sensitivity, better heart health, and possibly even reduced anxiety. Researchers suspect these outcomes are due in part to the polyphenols (such as flavonoids) in cocoa. There have been tomes upon tomes written on the supposed benefits of polyphenols—from red wine cures to green tea cleanses. But precisely how they work in the body has been a bit of a puzzle. Scientists were unsure how these compounds were actually absorbed and used by the body. "The perplexing things about these compounds is that they're not actually digested," says UCSF's Peter Turnbaugh. "We're eating polyphenols that are coming from all these different foods, and they're essentially trapped in the gut. And so it's been a great mystery." But, he says, "there's a very small fraction that gets broken down by gut bacteria into downstream metabolites that can then be absorbed." Indeed, one study by another group of researchers found that "cocoa-derived flavonoids modulate the human gut

*** The production of cocoa beans itself relies on a detailed fermentation process, but the microbes from that step are long gone by the time we receive the finished product.

microbiota toward a more 'health-promoting profile' by increasing the relative abundance of bifidobacteria and lactobacilli."

Of course, these benefits come from the cocoa itself, which, as we consume it, is often diluted by the less healthful sugars and fats in a chocolate bar. But perhaps the Swiss are onto something. They are the world's most zealous consumers of chocolate, eating an average of nearly twenty pounds per person each year (more than twice as much as Americans). Despite the local diet, heavy in rich foods, the Swiss have the third lowest rate of death from heart disease in the world and among the lowest rates of colon cancer in Europe.

• • • • •

Far from the alpine meadows of Switzerland, in a lab near the swamps of Louisiana, one gut is probing the inner secrets of cocoa. Here in Baton Rouge, researchers found that if they put cocoa powder in one end of an experimental gut fermenter, anti-inflammatory compounds came out the other.

Their gut contraption exposes cocoa to human digestive enzymes and then to gut microbes.††† These tests have turned out some of the most compelling results about the potentially prebiotic role of cocoa. In their experiments, human enzymes left behind cocoa flavonols (a type of plant compound linked to reduced occurrence of many chronic diseases). These flavonols were then fermented by the microbes into tiny anti-inflammatory compounds that the body could indeed absorb. Additionally, polyphenols found in the cocoa might also encourage the growth of more beneficial microbes along the digestive tract.

Not bad for a bean. With some help from our local microbes.

††† Harvested from graduate student "donations."

• COCOA OATMEAL •

Not every microbe-enriching meal needs to be a homegrown affair or requires a specialty-store scavenger hunt. Some can be as simple as a slight twist on your daily breakfast.

Food biochemist John Finley is the senior researcher behind the fermentive work at the Louisiana State University lab that uncovered some of the prebiotic capacity of cocoa. He says his findings inspired him to add it to his morning oatmeal. Although his wife looks askance at his new breakfast topping, he knows the addition is a favor to his microbes.

Try this simple and nutritious dish as breakfast or a snack. The ingredients include rolled oats, unsweetened cocoa powder, and optionally walnuts, cinnamon, and/or yogurt or kefir.

Bring 1 cup of water to a boil.

Add 1/2 cup rolled oats and simmer until done (10 to 20 minutes).

Transfer to a bowl and sprinkle 1 teaspoon (or more) unsweetened cocoa powder on top; optionally top with:

CHOPPED WALNUTS

CINNAMON

A DOLLOP OF YOGURT OR KEFIR
(WITH LIVE AND ACTIVE CULTURES, OF COURSE)

Other labs are creating even more dynamic and exacting methods to see what might be going on in the gut when we eat foods like cocoa. I visited one such lab in Switzerland to see the sloshing science for myself.

In a small room on the Swiss Federal Institute of Technology Zurich sit six guts. Swirling inside each is a brown liquid, which you can see because these guts are clear and built for study. Each is a glass cylinder, connected via an array of tubes. They are the feces-filled pride and joy of Christophe Lacroix, a food biotechnologist at the university's Institute of Food Science and Nutrition.

The room smells like, well, watery stool. Because that's essentially what is mixing—diluted from real human samples. It's an odor that nonchalantly plasters itself to the inside of your nostrils.‡‡‡ But the researcher on duty running her experiment doesn't seem to mind. And despite the biomaterial, the lab requires just moderate safety precautions. "It's no more dangerous that a public restroom," my fermenter room host Tòmas de Wouters tells me. Still, he says, "Don't touch anything you don't have to." (Also probably good advice for a public restroom.)

The full contraption replicates, if not the appearance, then at least the functions of the lower human GI tract—without having to send food through lengthy digestion chambers, as in other experimental models that have been built. The cylinders, like the human gut, are anaerobic. Microbes are extracted from donors' fecal samples, and then these microbes are cultivated and divided through the system and into the different gut chambers. This allows researchers to have five experimental "guts" that start with the same microbial population—in addition to one control gut. The scientists can manipulate the environment to regress the sample to resemble the population of microbes in the upper large intestine or compare what that might look like with parts of the lower large intestine. And it can run continuously for months at a time.

This elaborate if odiferous setup also allows researchers to see the metabolic products and ecological shifts with changing conditions. When I visit, they are testing various fibers to see how much is degraded by microbes and what short-chain fatty acids are produced. Specifically, they are investigating soluble fibers that encourage the microbiota to produce butyrate, which supports the growth of essential cells that line our large intestines. The rate of production of this beneficial compound, of course, also depends on the sample donors and their microbial

‡‡‡ Not a place, I learned the hard way, that you should make your last stop before a lunch meeting.

makeup, so the researchers run the same tests on samples from various donors to compare results and look for trends.

They can also study very specific forces as they relate to health, from retention times (which is the time between what you might have guessed) to pH shifts. These important variables can get muddled by myriad confounding factors in real-world human guts. "It is such a mess," says de Wouters, of all the dynamics at work in people living their lives out in the world. But in this very controlled laboratory environment they can see that changing only the pH—by perhaps just 0.2—makes a big difference in these guts. "You see it in a way you can't see it in donors," he notes. The researchers can also specifically test samples from people with IBD and other illnesses. Or they can develop unique microbe communities and test how they do.

As Christophe Lacroix and his colleagues noted in a paper, the standard genetic sequencing of our microbes has been a boon to figuring out *who* is in our guts. But as we've learned, it doesn't tell us anything about what they *do*. Now we need to start "unraveling the complexity of microbe-microbe interactions and identification of niches central to gut fermentation," they write.

And so he designed his glass gut fermenters. "We mimic in fermenters the functioning of the gut microbes," Lacroix explains in well-enunciated English when I visit him in his sunny office, not far from the fermenter room. "We can also cultivate in the gut fermenters to mimic an infection. We use, afterwards, these models to test for treatments like antibiotics, like probiotics."

Lacroix and his colleagues are keenly interested in prebiotics and their impact on the gut microbiome and its products. "Prebiotics are always important components, so they can be used especially in order to promote some activities—either metabolic or enhancing certain microbial populations," he says. "Now the challenge is to find prebiotics that are more specific and that are more targeting certain microbial populations.

This might require that the concept of prebiotics is maybe expanded toward a broader range of compounds." This move happened after we spoke, in fact, as scientists agreed that the term *prebiotic* would include any substance that helps microbes have a beneficial impact on their host—whether in the gut or not. Which makes sense, as we should be happy to help our beneficial microbes whether they are living in our guts or on our skin.

He and his colleagues can also look at diet more broadly. "Any food will impact on gut microbiota," Lacroix says. "With food, it's even more difficult because you have difficulty to control the intake of food. We know that some foods, like indigestible fiber, they feed the microbes—they change the balance of the metabolic activity. If you work with many foods, you have many components in these. And the components, how are they absorbed? How do they come into the contact of the microbes? How do they change the microbial features and activities? That's a big question." And so they are tackling it—six guts at a time.

In the meantime, I have to ask Lacroix what he has gleaned from all of this mechanical gut work—and if he has applied anything he has learned to his own life. He says he has maintained a balanced approach to his own diet. "For me, I've had more simple approach to food," he says. "Food should be good, should be healthy, and should be pleasant and diversified. This is the main point. We need a variety of food so that a variety of microbes can feed on the food the natural way."

• • • • •

It might seem like we have covered it all—from cocoa to kimchi to kefir. But there is still one realm of food that we have not yet touched. It is an area of digestive complexities and audacious products. It is the vastly unapologetic world of microbe-fermented meats.

CHAPTER EIGHT

The Undead
Meat

Why ferment meat? It's at once an obvious and a vexing question. If there is one type of food rot you would be especially wise to control, it would be meat rot. Vegetables might get slimy, dairy might get stinky, grains might get moldy, but expired meat can be especially, expeditiously hazardous if populated by the wrong microbes. Look no further than our innate repulsion to the smell of rotting flesh. Yet if you've had a cured sausage or a traditional fish sauce, you've had fermented meat. And you know that if it's done right, it can be delicious.

• • • • •

Throughout much of human history, meat has likely been the exceptional food rather than the foundational. Traditionally, its availability was limited, based on seasonal herd migrations, happenstance, or the result of time- and resource-intensive husbandry. That meant many groups found ways to make this precious source of protein and fats last beyond its natural shelf life. Which, save in arctic climates, isn't long.

So if vegetables become pickles, dairy transforms into yogurts and cheeses, and beans morph into pastes, what exactly does fermented meat *become*?

The answers to that question are many: prosciutto, fish sauce, pickled sushi, the notoriously powerful surströmming (fermented fish in a can), and kiviak (the hollowed-out seal stuffed with whole seabirds, closed up, and buried in the ground to ferment for months at a time). But before we get too deep in this wild world of fermented flesh, let's start with something a little more familiar: sausage.

Dry Rot

Not unlike cheeses, salami and other fermented meats go through an intentional, controlled, and microbially rich process of ripening. When prepared and handled in a traditional way, these meats can also offer a diverse array of potentially beneficial microbial life.

Unfortunately, this is not always the case with many cured meats we find on our plates these days. "In the United States most of the sausages and salami that we're familiar with are given a heat process to ensure food safety, and shelf life. This heating step, however, also inactivates most of the bacteria," says Robert Hutkins, fermented foods and human health researcher at the University of Nebraska–Lincoln. "In Europe, especially southern Europe, they consume a lot of uncooked dry fermented sausages—*those* have a lot of microorganisms."

There are a couple of basic types of meats cultured in this microbe-rich way. There are whole-cured foods such as prosciutto and jamón. These are full legs (in this case, of hogs) that are cleaned and then left in salt for several days to several weeks, allowing the salt (and sometimes extra applied weight) to draw out moisture. The legs are then rinsed of the salt and hung to dry for months to years—before being sold,* sliced, and enjoyed. One study of Iberian ham found species of *Micrococcus* and *Staphylococcus* (which are common on human and animal skin) were live and active in the meat and helped to develop its characteristic odor and flavor.

And then there are hard sausages. These aren't your Jimmy Dean patties or breakfast links. These are blended meats placed into casings and carefully aged. Unlike the sausages we might cook for brunch or grill up for a barbecue, these are eaten "raw." They are to true raw meat about what sauerkraut is to raw cabbage. But they are fundamentally transformed, simply not over heat. The fermentation process converts them into edible, safe, delicious foods.

Also like for sauerkraut, the process for making these sausages is all about creating the conditions to encourage the microbes you want to grow. Leo Miele, a food researcher specializing in fermented animal products at the Swiss Federal Institute of Technology Zurich says, "In certain parts of Europe, they do it sometimes still spontaneously." Meaning, let microbes come and grow as they please. "Meat, in principle, it's sterile. But in the slaughtering process, it has contact with the environment," Miele explains in his Swiss accent. "You have a kind of input of bacteria. The lactic acid bacteria help to ferment the meat, and the *Staphylococcus* species contribute to the aroma. That's a typical meat fermentation performed everywhere in Europe."

• • • • •

Sausage can be made out of any type of lean meat—including pig, cow, goat, sheep, and even camel or horse. Regardless of the base

---

* Sometimes at prices, for the finest jamón ibérico, upwards of thousands of dollars per leg.

meat, added fat usually comes from pork. To make this type of fermented sausage, the lean meat and fat are chopped and blended and combined with salts, seasonings, and often, curing nitrite (which acts as a preservative and also lends cured meats their pink-red hue). The meat mush is then stuffed into casings—traditionally made from animal intestines. Any air bubbles are pushed out. Lactic acid bacteria that enter the meat during processing slowly turn carbohydrates (from the meat and sometimes from added sugar) into lactic acid, making the environment more acidic and less hospitable to would-be spoiling microbes.

Historically in Europe, this aging process took place in the cooler months, which provided the lower ambient temperatures for a slow and steady fermentation that also discouraged unwanted microbial growth. Additionally, the humidity level is crucial. Unlike many of the ferments we have encountered that rely on submersion to create an anaerobic environment, sausage is hung in the open air. But an overly dry environment can lead to the sausages drying too fast, resulting in undesired cracks in the sausage. The process takes anywhere from a few weeks to a few months. Fermented properly, these longer-fermented hard sausages should be able to last at room temperature and in moderate humidity for more than a year. Not bad for raw meat.

So how does it work? Lactic acid bacteria lower sausage pH to around 5 or 5.5. Predominant bacteria include species from the genera *Lactobacillus, Pediococcus, Staphyloccus,* and *Streptococcus*. One study of traditionally fermented sausages made in northeast Italy found some 150 strains of lactic acid bacteria, the majority of which were either *Lactobacillus curvatus* (strains of this are sold as a probiotic and might help lower cholesterol) or *Lactobacillus sakei* (which might help the immune system).

Larger commercial operations usually decline to rely on chance and instead use a starter culture, gaining more reliable and controllable (if less complex) results. As is true in dairy products, various blends of strains can produce different flavors and textures. "*Lactobacillus* results in fast acidification," the UN Food and Agriculture Organization notes in one report on sausage making. "*Pediococcus* leads to slower and milder acidification. Selected *Staphylococcus* strains cause a speedy reduction of nitrite, stable curing color, and reduced risk of fat rancidity." The blend might also depend on the size of the sausage, with smaller sausages (those up to about two and a half inches in diameter) generally taking a fifty-fifty blend of *Lactobacillus* and *Staphylococcus* strains, and larger sausages (up to about four inches in diameter) often using more *Staphylococcus*, which help with the longer curing process.

It might seem like bacteria have the process covered, but they are not the only microbes at work. The exterior of the casings often gets covered with mold, which can occur spontaneously or be intentionally introduced. These desirable molds contribute part of the classic taste while also discouraging the growth of unwanted surface microbes. A study of Turkish dry-fermented sausage known as sucuk found 255 strains of yeasts, including *Debaryomyces hansenii* (which loves salty environments so much it is found in extreme places such as the Great Salt Lake in Utah, as well as in many cheeses).

Shorter-fermented, softer sausages are known as semidry or summer sausage. A *Staphylococcus*-heavy starter helps rapidly reduce the pH and encourage the activity of partner-in-fermentation *Lactobacillus* species. These sausages might be cold-smoked as well, but generally they are aged only a matter of days to weeks before they are considered shelf stable and ready to eat. They are sourer than long-aged sausage (with a pH of around 4.8 to 5.4) and quite a bit moister. Some are even spreadable.

• • • • •

People living closer to the equator have their own fermented sausage traditions. Only these are much faster—and damper. In Vietnam, nem chua is a sour sausage appetizer made of raw pork, pork rind, and ground rice, along with spices and salt. Unlike the air-aged sausages of Europe, this version is wrapped in leaves (generally banana) to create an anaerobic environment and left to ferment over a day or two. The added ground rice provides a starch for lactic acid bacteria to consume, hastening fermentation. In its cured state, it contains numerous microbes, including, among others, *Lactobacillus plantarum, L. brevis* (strains of which are sold as probiotics), and *L. farciminis* (another probiotic, studied for its anti-inflammatory capabilities). Thai cuisine has a similar, short-fermented sausage. Called naem, it also contains *L. brevis* and *L. plantarum*, as well as other bacteria and yeasts.

Saucy

On the other end of the fermented meat spectrum, many animal protein preservation methods turn not to drying but instead to liquefying.

In the tropical coast-lined countries of Southeast Asia, for example, fish sauce has been an age-old answer to the challenge of preserving plentiful catches. Classic Vietnamese fish sauce, known as nuóc mám, in its purest form contains just anchovies and salt. The fish are mixed with large amounts of salt (roughly three parts fish to one part salt) and allowed to ferment in wooden barrels for a year or more at steamy temperatures of 95 to 122 degrees. Once adequately acidic, the blend is pressed or drained out and filtered. It's used locally as seasoning for a variety of dishes, to which it brings a dash of *Lactobacillus plantarum, Bacillus subtilis*, and several other beneficial microbe species. And it finds its way into many bowls, with Vietnam, a country of 89 million people, producing some 58 million gallons of nuóc mám a year. And this product is not exclusive to Vietnam. This sort of sauce has been developed in numerous countries. A similar Thai version is known as nam pla; the Japanese make shottsuru; and in the Philippines, the local fish sauce is known as patis.

Many other fish sauces use a variety of seafood and processes. "They're pretty intense," U.S.-born Tokyo cook and author Elizabeth Andoh says of many of these fermented fish sauces. A bit more potent than the version you might buy at your local grocery store. Some use all matter of small catches—diminutive fish, eel, shrimp, squid, octopus. Some dry the catch first. Some use the whole animals, others just the guts. All of them create a rich, umami, salty sauce that contains a stew of nutrients and microbes. One study cultured microbes from four types of Asian fish sauces and found nearly forty different microbe species, including bacteria as well as fungi. A study of Thai fish ferments identified more than a dozen species of lactic acid bacteria, including two entirely new species. And this study was conducted without genetic sequencing, which likely would have picked up far more species and strains.

Although fermented fish sauce today is primarily associated with Asian cuisines, it was once a more common condiment in Europe as well. Ancient Romans added flavor to their dishes with garum, a recipe they adapted from the Greeks. Similar to the process still employed elsewhere around the globe, it was made by blending fish innards and salt and letting the mixture ferment. One small town on Italy's Amalfi Coast still produces a version, known as colatura di alici, which translates to something like "leakage of anchovies." Buon appetito!

• • • • •

Many meat preservation techniques create a thicker, even more potent paste. Mám chua is Vietnamese fermented fish paste that comes in several varieties and includes, among other species,

Lactobacillus farciminis (a strain of which has been patented to treat digestive issues), *Staphylococcus hominis* (common on animal skin—and also produces some of the body odor scents on humans).† The Cambodian version is known as prahok, which is made by crushing cleaned fish (traditionally by foot), salting them, and then fermenting them for a few weeks . . . to a few years.

A particularly effluvious paste is the Japanese shiokara, which is composed of various small sea creatures and viscera, fermented with salt and rice. "Microorganisms contributed a lot of the particular odor of shiokara," writes a team of Tokyo researchers, who found *Staphylococcus* to be responsible for much for the acid creation in the gooey dish. Each type features a different animal as its main ingredient (such as squid, cuttlefish, or sea urchin roe). Even scientific papers, known for their dry technical discourse, leak words such as *unique, strong,* and *potent* into their titles. To give you another indication of its strength, some locals chase a bite of shiokara with whiskey—rather than the other way around.

Submersion

There are plenty of other ways of fermenting meat than drying it out or liquefying it. You might consider these the pickles of the meat world.

Now, of course, meat sitting around for long periods of time is a potential feast for the dreaded *Clostridium botulinum*, the bacterium that creates the severe neurotoxin that causes botulism. The disease was in fact long known as "sausage poisoning," and its name comes from the

† Our sweat itself is odorless; we have our microbes to thank for fermenting it into malodorousness.

Latin word for sausage, *botulus*.[‡] The bacterium itself was first isolated in the late nineteenth century after an outbreak of illness in a Belgian village stemming from a smoked ham.[§] Although *C. botulinum*'s native habitat is soil, it can easily find its way into food—animal or plant—and multiply there, even tolerating life without oxygen. Highly acidic environments, however, are deadly to the bacterium—hence the safety of ferments.

To avoid this fate, people have added meats to salt-brine pickling mediums alongside veggies, such as sauerkraut or kimchi. Beyond a saltwater ferment, another option is to ferment meats in sour milk. Icelanders have traditionally used this method to ensure hardy nourishment through the winter. One of these wintertime treats is *súrir hrútspungar*. If you think the name is difficult to digest, just wait for the dish: ram's testicles stuffed into a sheep's stomach fermented in whey. One Icelander says that "when it is done, it gets like a jelly. This testicles are really soft and sour!" Perhaps some simple súr hvalur (whale fat preserved in sour milk) sounds more appetizing.

• • • • •

In Japan, there are nuka (rice bran) fish pickles, made like the other quick pickles in the nuka pot. But there is also a slightly more daring dish, known as narezushi, which is thought to actually be the origin of modern (unpickled) sushi. Said to have been first created in Southeast Asia before arriving in China and then Japan more than twelve hundred years ago, narezushi was one way of preserving fish catches. Fish were cleaned, salted, and nestled into a vessel of fermenting rice. As the story

‡ Referring to its origin, rather than its shape.

§ Prepared for of all things, a funeral dinner.

goes, people eventually started eating the rice, too, and sushi, of sorts, was born. Fresh fish sushi became more popular by the Edo period in the seventeenth century. But with its rise, diners began missing out on a trusty source of microbial diversity. One genetic study of six different narezushi products, for example, found more than seven hundred different strains of bacteria, most commonly belonging to the *Lactobacillus* or *Pediococcus* genera. Another study found the most abundant species overall to be *Lactobacillus sakei* (a species that might help the body ward off infection).

An even more pungent Japanese pickled fish is kusaya, which is often made from mackerel scad on islands south of Tokyo. The fish is pickled in a low-salt, microbe-filled brine for up to a day and then dried. The brine—sometimes referred to in this case as a gravy—is often kept for years (and sometimes generations), developing a honed microbial community. This "stinking fish" is referred to as an acquired taste.

A similar dish exists in Turkey and Greece. Called bonito fish (not to be confused with the smoked and aged bonito fish flakes of Japan), it is cleaned and placed in a brine for one to a few days. After brining, it is salted and packed into a container for weeks, creating lakerda. This fish can then be eaten right away or kept for later consumption by submerging it in olive oil. It's commonly served raw as an appetizer with slices of onion, adding prebiotics to the microbe-infused snack.

Odds and Ends in the Forge of the Microbe

Many countries in Africa, such as Sudan, have a deep history of animal-product fermentation. Hamid Dirar, author of *The Indigenous Fermented Food of the Sudan* and a researcher at the University of Khartoum there, spent years collecting and studying the fermented foods of that country. He gathered much of his information by speaking with local women, the primary fermenters, who learned these crafts from their mothers and grandmothers. "The Sudanese seem to bring just about anything edible or barely edible to the forge of the microbe, to the extent that one could confidently say: food in Sudan is fermented," he writes in a report for the U.S. National Research Council. And he is not exaggerating. Of the country's fermented foods, he enumerates: "bones, hides, skins, hooves, gallbladders, fat, intestines, caterpillars, locusts, frogs, and cow urine."¶

There is a sausage stuffed with fat and hung to dry, known as skin. There are the sun-dried, pounded balls composed of internal animal organs that are slowly fermented, called twini-digla. A more immersive method is used for beirta, which is made specifically of male goat meat: "small pieces of muscle meat, lungs, kidneys, liver, heart, etc., are mixed with milk and salt, packed into a clay pot, and allowed to undergo some sort of pickling, presumably," Dirar offers.

Tougher parts of the animals are made edible by means of microbial action. Animal skin and hooves are buried in mud or wet ash to ferment. "Fresh bones may be fermented in a number of ways," Dirar writes. "Large bones with pieces of attached meat and tendons, may simply be thrown on a thatched roof to ferment slowly for weeks or even months to give a product called *adum*," he notes. "The meshy ball bone endings of the ball and socket joints may be pounded fresh and fermented into a paste called *dodery*. The vertebrae of the backbone may be chopped into smaller pieces that are sun dried, pounded with stones, mixed with a little water and salt, molded into a ball, and allowed to ferment," the final product being kaidu-digla, the unapologetic name meaning "bone ball."

Rather than slurp down a whole fermented hoof stew, however,

¶ Unfortunately, he does not disclose much in the way of detail about these few latter products.

Dirar notes that most of these items are used as nutrient-rich condiments added to a porridge of sorghum or millet, which create a native microbe-feeding, fiber-rich basis for the meal.

Taking It Underground

Perhaps some of the most common strategies for preserving animal flesh involves not brining it or saucing it or pasting it or pounding it, but burying it. Some are buried whole, as in the case of Korea's infamous fermented skate hongeo-hoe. Others are fermented *inside* other animals: the legendary birds-in-a-seal-carcass kiviak.

Before we leap into these delicacies, let us start with something simple, like central Norwegian rakfisk ("moist fish").

This fish preparation appears in historical documents by name as early as 1348, and it was likely being prepared well before that. In central Norway several hundred years ago, salt was difficult to come by. So freshwater fish are only very lightly salted before being packed tightly into barrels and kept underground for anywhere from three months to a year. Being submerged in a brine of their own juices, the fish undergo spontaneous fermentation, with lactobacilli (particularly *Lactobacillus sakei*, the immune-system helper) being the most prevalent bacteria.

Rakfisk was typically eaten in the late fall and midwinter, when summer-caught fish were fermented and ready. What is it like? It has a "characteristic taste, odor, and somewhat spreadable texture," a Scandinavian research team notes, adding that, "the taste, smell, and spreadability increase with the ripening time."** It is commonly served with bread, onions or leeks, and sour cream.

** .The ubiquitous polite use of the word characteristic in academic writing to describe many of these

The perhaps more familiar Swedish gravlax ("buried salmon") was historically made like rakfisk, but now it is commonly just marinated for a few days and is not shelf stable. Even at Ikea.

• • • • •

Some food items that are nominally shelf stable are really a bit less so than advertised. As we well know, bulging cans in the grocery store or pantry are generally to be avoided (for fear of contamination with *Clostridium botulinum*). But in the case of our next subject, an expanding can is a sign that something safe, good, and—to some people—delicious is going on inside.

Sweden's notoriously potent surströmming continues its fermentation even after canning. And its name, meaning "sour herring," if you have ever encountered it in person, might seem like the understatement of the culinary world.

The creation of surströmming has a very precise seasonal rhythm to it. Baltic herring are typically caught just before they spawn—May through early July. The herring are immersed in a salty brine for a day or two before being removed for cleaning. They are then placed in wooden barrels, again with low salt levels, and sealed for a few weeks of cool northern latitude summertime fermentation. After this segment of culturing, between the beginning of July and August, the fish and brine are placed into cans and sold to distributors in mid-August (traditionally not to be sold to consumers before the third Thursday of the month). After about six months, microbes fermenting away inside the can have produced enough carbon dioxide to bulge out the can's sides, creating its signature rounded shape. Before canning technology arrived in the

---

more potent ferments verges on the pathological. It is also a little unclear in what direction, exactly, the taste is increasing. I will hazard to guess it is the "characteristic" direction.

1800s, the fish were kept in the large barrels and transferred to smaller kegs for home consumption. They were doled out as Swedish army rations starting in the 1600s—a kind of SPAM before there was SPAM.

What is happening during this fermentive process? Early on, lactobacilli dominate. "The origin of these is most probably the barrels," notes a team of Scandinavian researchers. "When using sterilized vessels there is no development of the typical surströmming flavor" (considered a negative outcome by those in charge). But these aren't the only bacteria at work. One of the principal types of bacteria behind the in-can fermentation is the *Halanaerobium* genus. "These bacteria produce carbon dioxide and a number of the compounds that account for the unique odor," report the researchers politely. How would they describe, scientifically, the food's full, fermented flavor profile? "Pungent (propionic acid), rotten-egg (hydrogen sulfide), rancid-butter (butyric acid), and vinegary (acetic acid)."

Despite this new scientific understanding of the fermentation process, bulging cans in the supermarket are still a little suspicious. To set consumers' (and their own) minds at ease, the Swedish Food Agency tested to see if various foodborne toxins could thrive in surströmming's unconventional environments. So they added microbes known to cause food poisoning to experimental batches of fermenting surströmming. After fermenting the fish according to its standard process, none of the introduced harmful strains grew. The beneficial fermenting microbes chased away the bad actors, demonstrating once again, the accumulated wisdom hiding in many of these ancient foods.

But just knowing this doesn't necessarily make the fish easy to eat. Many surströmming devotees defend the food, explaining that its olfactory bark is worse than its bite. But it is so smelly that cans are often opened outdoors, even at dedicated surströmming parties. Some airlines actually ban it, considering the already pressurized can to be in danger of exploding. Aside from the physical force, it might create an emergency of a different flavor in the cabin.

• • • • •

Even beyond bulging cans of fermented fish, there are still some even more challenging—and microbially fascinating—fermented meats out there. Like alkaline-fermented poisonous shark.

Hákarl, an infamously malodorous Icelandic food, is made from Greenland shark, which is toxic to eat if not properly prepared.[††] Icelanders might have been making this food for some seven hundred years. It also evolved without salt. You might think that strange, for a nation surrounded by salty seas. But for hundreds of years, crystallized salt was difficult to create in the cold without using copious amounts of precious firewood. So hákarl cleverly mixes fermentation with drying as a double means of preservation—and a double guard against harmful microbes.

To make hákarl in the traditional fashion, the shark is first cleaned and cut into large chunks. These sections are placed into a gravel pit near the ocean. They are then covered with gravel, seaweed, and rocks, the weight of which help to press out moisture. (A more modern method involves fermenting pieces of the shark outside in large containers.) There the meat ferments for a month and a half to three months. Once it has fermented properly, the meat is dug up, cut up, and hung up to dry for weeks to months.

During fermentation, something interesting is happening, something very different from our other, acid-based meat ferments. In this case, bacteria break down high amounts of urea in the meat, creating substantial quantities of ammonia. If you thought some of the lactic acid bacteria ferments were intense—say, a strong kimchi or even a Thai

---

†† These sharks' bodies contain high amounts of urea as well as of trimethylamine N-oxide (TMAO), a harmful compound that can be poisonous to humans. These sharks might also be a little gamy, given that are the longest-lived vertebrates on the planet, with one specimen clocking in at nearly four hundred years—give or take a century or two. You know, just two hundred years or so before canning was invented.

sausage—brace yourself for alkaline-fermented shark. Which, to put it mildly, can take some getting used to. The pH of the meat goes from a slightly acidic 6 to a highly alkaline 9 in the final product. Hákarl is typically served in small pieces. It is often chased with a shot of local liquor (it speaks volumes that, as with the Japanese seafood paste shiokara, straight alcohol is not the chased but the chaser).

So we all have an affinity for at least one perhaps peculiar food we grew up with, but hákarl is yet one more dish that begs us to ask the lingering, odiferous question on so many of our Pasteurian minds: How could such strong foods have gained a foothold in local culture in the first place? "There is good reason to believe that the general sensory acceptance of aroma and taste was quite different in former times," writes the team of Scandinavian researchers. In a world where refrigeration and canning were unknown, and harvests and catches were often fleeting, food was eaten on a continuum of freshness. So "the occurrence of rancid and putrid aroma was a much more natural part of the daily food," they note. And of course, in the case of hákarl, "the fermentation of shark renders a potentially harmful raw material (fresh shark) into a nutritious food product; in such a way that the final ready-to-eat product can be stored for long periods (up to several years), without spoilage." So it seems that now *we* are really the spoiled ones.

• • • • •

Alkaline seafood fermentation is not reserved for Northern Europe. Korea boasts an intensely ammonia hongeo-hoe. To make this delicacy, whole skates are packed in straw in an earthenware vessel and left to ferment. No salting, no brining, no mercy. It is sliced for serving and eaten in all of its raw glory. It is sometimes served with kimchi or the alcoholic fermented rice drink makgeolli.

Amazingly, this preparation has emerged elsewhere as well. "Skate is traditionally processed in Iceland in a similar manner," writes the group of Scandinavian scientists, "by placing it in a pile and allowing it to cure or ferment for a period of a few days to few weeks." This ferment carries its own special microbial population not found in very many other foods. These microbes include, among others, species from the genera *Aliivibrio, Oceanisphaera,* and *Photobacterium* (some species of which are bioluminescent)—if you are looking to *really* spice up the genetic environment of your gut.

• • • • •

And lastly, perhaps one of the most impressive meat fermentation tours de force is our favorite, kiviak (or kiviaq), the infamous flock of birds fermented in the body of a seal. This delicacy is rumored to have finally felled the notoriously hardy Danish Arctic explorer with the exceptionally hardy name of Knud Rasmussen.

Native Greenland Inuit groups would begin to prepare this wintertime dish in the spring. In that season, little auks (small black and white seabirds) were numerous. These birds have been an important food source, particularly in northwest Greenland, for upward of eight hundred years. And evidence suggests nets have long been used to catch large numbers of these birds at a time. But what to do with so many birds? And what about making it through those long Greenland winters? Why, ferment them, of course. No fermenting vessel handy? Grab a seal.

To make this resourceful fermented food, hundreds of these birds are caught and then stuffed whole into a hollowed-out sealskin. A layer of fat from the seal surrounds the birds, and as much air as possible is squeezed out, creating an essentially anaerobic environment, like any pickling crock. Locals then bury the seal-sack of birds for anywhere from months to a year and a half. The result is preserved meat that is entirely edible from head to toe (save for the feathers). The birds are consumed for special occasions—and often out of doors (perhaps for a similar reason as surströmming). Not going to be in Greenland any time soon to try it out in person? Here's a sample: The flavor of kiviak has been compared to "a mixture of vinegar and the strongest imaginable blue cheese."

It wasn't the smell or the taste of kiviak, though, that did in Rasmussen, the Danish explorer. Having spent much of his childhood and his adult expeditions in Greenland, he was surely familiar with the food. Alas, he died from pneumonia, which he was rumored to have caught only after being weakened by a case of food poisoning brought on by these fermented birds. But if prepared properly, they are safe to eat. Unfortunately, little microbial research has been done to elucidate the hidden world within that sealskin.

The Other Side of Meat Fermentation

Despite the many potentially beneficial microbes lurking in these fermented meats, meat products that are fermented *inside* of us have different consequences for the gut microbiota. Looking at the actual prevalence of meat in many traditional diets, we see that meat was much more of a sometimes-food—with plant-based foods as the basis. So it is interesting and perhaps telling that meat has a very different impact on our own gut microbiomes—and consequently our bodies.

In the diet microbiome study conducted at Harvard University discussed in the Introduction, ten volunteers first ate an all-animal-product diet and then a high-fiber all-plant diet (or vice versa). During the five-day dietary changes, the largest alterations came from the animal-product group. Their microbes shifted in population and genetic action to be more tolerant of bile, which the body produces to help break down meats. Particularly striking was the increase of *Bilophila wadsworthia* in subjects on the animal-product diet, which proliferated within a matter of days. (This even occurred in the lifelong vegan who volunteered to eat all animal products for this study.) *Bilophila wadsworthia* has previously been linked to high-fat diets and to inflammatory bowel disease. Other research has shown that high-fat diets can decrease the amount of bifidobacteria, which are big contributors of beneficial short-chain fatty acids that help keep our gut lining intact, an important step in limiting inflammation.

Of course, animal products are wildly different and contain many different components: fats, proteins, etc. And each of these dietary aspects can alter the microbial gut landscape in different ways.

Animal fats are powerful actors in this respect. But not all fats have the same effect. Fat from fish and fat from pork behave quite differently in the body on many levels. We know from decades of studies that fish oils and fatty fishes are linked to numerous beneficial health outcomes. And research is parsing how the microbes in our guts might be helping to determine these different outcomes. In one study, researchers fed mice diets in which their fat came from either lard or fish oil. After eleven weeks, the mice getting their fat from lard showed predictable patterns of metabolic disease, including increased inflammation. The two groups of mice also developed quite different microbe communities. Those on the lard-heavy diet had higher levels of *Bilophila* (which is found in high concentrations in people with inflammatory bowel disease) and *Turicibacter* (which has also been associated with IBD). The mice eating fish oil had more Actinobacteria, Verrucomicrobia (a phylum that includes *Akkermansia*), and various types of lactic acid bacteria.

Could it be, though, that the poor health conditions were impacting the microbe patterns, rather than the other way around? Interestingly, when researchers fed mice lacking microbiotas on the same two diets, those on the lard diet showed less inflammation than their microbe-colonized peers on the same diet. This indicates that microbes likely play a substantial role in how the body responds—for better or for worse—to different types of fat.

As a final twist, the researchers working on the same study transferred the gut microbes of the lard- or fish-oil-fed mice into other mice who had had their bugs mostly wiped out by antibiotics. The newly colonized mice who received the fish oil microbe profile stayed leaner when fed on lard than those that received the lard-trained microbiotas. This finding suggests that not only are our microbes malleable by our diet, but also that they might help steer us toward healthier outcomes even when treated to the occasionally unhealthful meal.

• • • • •

We have also seen how the foods we feed our microbes impact the things that they feed us in return. So just as prebiotic plant fibers enable the gut bacteria to produce beneficial compounds such as short-chain fatty acids, other foods can spur the microbes to make less helpful—sometimes even harmful—products. This is a reminder that just because something is made by the microbiome, perhaps even by a "good" microbe, that doesn't mean that it is good for us.

In a Western diet, with comparatively fewer whole grains and veggies and relatively more refined carbohydrates and animal products, there often isn't a large enough amount of prebiotic fiber to go around. So most of the complex carbohydrate fermentation occurs in the upper regions of the large intestine (also known as the proximal colon). These favored substances get depleted as the food (or the digesta, as it's called at this point) moves along the lower intestine. Thus, the local, lower microbes must find other things to ferment, which in the case of the Western diet, is "notably protein or amino acids," note the authors of one paper in the journal *Gut*.

Unfortunately for many of us, "fermentation of amino acids

produces a range of potentially harmful compounds," the researchers write. "Some of these may play a role in gut diseases such as colon cancer or IBD." These amino acids are found in high concentrations in meat products—a large component of typical Western diets but infrequent in many traditional and longevity-linked cuisines. And just to bring it full circle, they note, "dietary fiber or intake of plant-based foods appears to inhibit this, highlighting the importance of maintaining gut microbiome carbohydrate fermentation." In other words: eat your vegetables (and legumes and whole grains).

We are still learning about many of these products that our gut microbes make. Recent science has shown that many of these microbe-made compounds can be quite potent. "There's a spectrum of drug-like compounds that the microbiota is producing," says Stanford University's Justin Sonnenburg. Lab-developed pharmaceuticals are rigorously tested. But these microbe-generated compounds are not. Nevertheless, they are "absorbed into our bloodstream and eventually metabolized, excreted by the kidney, or sent back out to the gut," Justin Sonnenburg says. "All these things are present in our blood right now, and they're different in you and me. They change with our meals." And some of them might not be so good for us.

One of the potentially harmful compounds is trimethylamine N-oxide (known more succinctly as TMAO), which has been implicated in heart disease. This substance originates with compounds found in red meat (carnitine), as well as in eggs and soy (lecithin). And it is present in much lower amounts in vegetarians and vegans. More than a century ago, probiotics pioneer Élie Metchnikoff maintained that a meat-heavy diet could exacerbate harmful microbial compounds. He seems to have been on to something.

Chinese microbiologist Liping Zhao honors this tradition and this scientific knowledge in his own diet. "I'm not a vegetarian, but I eat a plant-based diet mostly," he says. "Occasionally I take some high-quality animal product, but not much. I'm not against animal food, but the key is to take the animal food to a level that you can digest and absorb everything. If you leave undigested animal food to gut microbiota, you feed the bad ones or you change the digestive system to favor the pathogenic ones to thrive. Animal food is hard to get throughout evolution—so I don't think, evolutionarily, we set upper limit for taking animal food. And also they taste good. It's very hard to eat just a little bit. When you eat, you often eat a lot because it makes you feel happy. But, unfortunately, too much animal food indeed changes the gut environment, change the microbiota, and then cause health problems. More and more scientific evidence is coming to show that. So, we really need a way to keep a balance to our diet where animal food is only a modest part of the system, not a very, very big part."

But what of our long human history as hunters? A hallmark of our evolutionary success seems tied to an abundance of protein—especially animal protein. There is a lingering image of the original human dinner as large game cooked over a fire. This fit with the mid-century American ideal of meat as the centerpiece to every meal—and still pops up today in the Paleolithic and other contemporary diets. But that is not quite accurate, says Stanford's Erica Sonnenburg. "I think the idea of going back to a diet that's more ancestral is good," she says. "But I feel like a lot of the actual nuts and bolts of it gets thrown out." Or at least the nuts of it. And the roots. And the fruits and leaves and seeds.

Observing the few groups that still make their dietary living without farming or animal husbandry—as we all did for 95 percent of human history—we see that meat is not the rule, but the exception. The caveman or Paleo profile was probably much more omnivore than carnivore. In fact, only the Inuit who live in the Arctic Circle and perhaps a very few others have traditionally subsisted on large amounts of animal protein, and their microbiota are likely adapted for this.

One group that has been studied extensively for their traditional gut microbiota are the Hadza in Tanzania. They are one of the last remaining groups that continues to get by solely on a hunter-gatherer diet. But *hunter-gatherer* has long been a misnomer,[‡‡] Erica Sonnenburg points out. "Really, they're gatherer-hunters," she says. Although they do hunt, much of their diet comes from foraged roots, berries, and honey. They consume an average of 100 to 150 grams of fiber per day. That's close to ten times the amount the average American consumes (and three to five times the daily intake recommended by the U.S. Food and Drug Administration), So perhaps it comes as no surprise that they also have a much more diverse microbiome. "If you look," she says, "it's the plants that are the majority of their diet, and the meat is like filler here and there when they have access to that. There are many days that they're vegetarian out of the fact that they just don't pull any game in." So it was for the human microbiome, since the days of our earliest ancestors. (Indeed, all of our closest living animal relatives also get by on plant-heavy diets. Chimpanzees eat a diet that is more than 95 percent plant-based, consuming food from more than one hundred different species of plants.) And so it still is, as so many microbiome researchers note, in most traditional diets still in existence.

• • • • •

Our study of the human microbiome is still in its very early stages. And although we still don't yet have a clear map of all of the details, we are learning that microbes are powerful mediators of our diet and our health. One day we might be able to closely monitor these

‡‡ Perhaps a vainglorious one that flatters ourselves as a species of born conquerors.

compounds to better understand our health profile and risk for certain diseases before they emerge. Such detection might allow for a window of opportunity to intervene and head off these harbingers of poor health. Perhaps with little more than a shift in what we're eating.

In the meantime, let's take what we have learned and feed our own microbes just a little bit better.

CHAPTER NINE

Bringing It Home

In the United States, live fermented foods have not been part of the mainstream culture. We've picked and chosen and invented our way to an American cuisine, but fermented foods never really made the mainstream cut beyond pasteurized yogurt. They have, however, been bubbling away in immigrant communities and small health food movements, quietly burbling in crocks and jars, pots and vessels. In recent decades, these foods, borrowed from other cultures and given a local spin, have come out of the culinary shadows and are now hard to miss at farmer's markets, upscale food shops—and even standard supermarkets across the country.

Now I can buy kombucha at the same place I get my garden hoses.

The sudden popular interest in these once exotic, sometimes challenging foods is thrilling, both healthwise and culinarily speaking. A spicy kimchi or a vinegary kombucha might have been a harder sell ten years ago, before we started to get a glimmer of the complexities and power of the mighty microbiome—and a taste for the enchanting flavors and textures these microbes can produce.

But in this explosion of popularity and commercialization, are there some things being lost? Traditions? Contexts? Microbes?

To find out, I traveled to Berkeley, California, to meet one woman who is still more focused on the fermentation process than on fad profits.

Nurturing Ecosystems

Just a few blocks from the Berkeley waterfront, in a light industrial area, a hotbed of modern-day American fermentation is quietly (okay, noisily) moving the field forward, creating dozens of unique cultured products for an eager local customer base.

At the Cultured Pickle Shop, cofounder Alex Hozven and her colleagues let their imaginations run wild for their small-batch operation, creating innovative takes on fermented classics, whether that's kombucha or kimchi or kasuzuke.

Inside, the small space is bright and abuzz with activity—chopping, jarring, and 1990s music playing over a sound system. Despite the open workspace, visitors are welcome, and a glass-doored refrigerator by the entrance offers that week's selection of foods for purchase. Which is handy, because Hozven isn't much interested in getting into the big supermarkets. "I don't really want to be in that race. What we're trying to do is go more hyper-local instead," says Hozven, who has a firm handshake and a simultaneously laid-back and intense demeanor. Their distribution chain spreads from their informal shop to area farmer's markets, and not much farther. But they're by no means small scale. By midday one Friday in May when I visit, they have already processed about 500 pounds of cabbage destined to be sauerkraut.

Sauerkraut is just one of the many offerings to which they give a keen eye—and a number of inventive twists. But where their process and attentions really shine is their kombucha.

Although the basic principles and processes behind kombucha are simple, Hozven says, "kombucha is a really, really, really challenging thing to do well." The direction kombucha takes "is here to sour." And within that, "you're looking for balance," she says. "There's a lot of subtlety."

• • • • •

The creation of kombucha is an unexpected sight for those who are used to seeing the clean, sparkling, colorful bottles (with fanciful names like Guava Goddess and Cayenne Cleanse) in a line of refrigerated shelves at the grocery store. Its Chinese name will start to give you a hint: it means something like "red tea fungus."

The beverage we think of as kombucha likely arose in China, perhaps more than two thousand years ago. In the interim, it gained popularity in Russia, before spreading to Western Europe during World War I. It lost favor there during the tea and sugar scarcities of World War II, but it regained steam and charged west with the health food movements of the mid- and late-twentieth century. There is still a bit of confusion surrounding the English name we use for it today.* But etymology aside, business is booming, currently bringing in some $400 million annually

* For instance, don't confuse it with another liquid from East Asia. In Japan, kombucha is a nonfermented broth made from kombu, a type of kelp. In Japan, what we call kombucha is known as kocha kinoko (as in Chinese, meaning something like "black tea mushroom"). So how did this fermented beverage we now call kombucha get its name? Some circulated tales report a physician named Kombu treating a Japanese emperor with a tea of some sort. But its early history remains murky.

in the United States alone—and growing by leaps and bounds each year.†

Kombucha gains its flavor and microbial power from a tightly knit community of bacteria and yeasts (its SCOBY, symbiotic culture of bacteria and yeasts). This SCOBY is far different from that used to make kefir. These kombucha-making organisms live in a rubbery union atop the liquid (scientists call this a biofilm, which is maybe not the most appetizing way to imagine a beverage's genesis). This takes the form of a brownish-grayish mat that floats on top of the liquid. The first steps to making kombucha, though, starts innocuously enough with sweetened brewed tea. Then add a bit of fermented kombucha (a little backslopping) to hurry along the process and drop the pH, before introducing the kombucha SCOBY (often also called a mother). Fermenting kombucha jars lined up with their large, irregular SCOBYs in them can look a bit like curiosity displays of dredged-up sea creatures suspended in formaldehyde.

The batch is then left to ferment anywhere from several days to a couple of weeks. The final product is a slightly vinegary, slightly alcoholic, effervescent beverage. Once the kombucha is ready, the SCOBY is removed and transferred to a new batch to begin the fermentation process all over again, adding to its thickness and layers with each batch it ferments. As these strange amalgams continue to grow, they are often divided and passed on to other kombucha brewers. The SCOBY, being alive, is constantly changing as it encounters different teas, environments, and handling. So even having been brewed from the same mother, no two kombuchas will be precisely the same.

It is also possible to grow your own SCOBY as long as you have access to some fermented kombucha to use as a starter. The little bits

† Some kombucha breweries are opening their own storefront "taprooms." And growler refills—once reserved for microbrew beers—are an increasingly common sight for kombucha.

settled at the bottom of store-bought kombucha bottles are often clumps of yeasts and bacteria. If fermented correctly, the SCOBY will assemble, grow, and float up to cover the top of the liquid. This habit also allows the oxygen-tolerant microbes in the matrix access to air, while helping to keep molds and undesirable microbes out of the liquid below.

The yeasts and bacteria in kombucha SCOBYs have a nice reciprocal relationship. The yeasts turn the sugars into glucose and fructose, while putting off alcohol and carbon dioxide. The acetic acid bacteria then turn the glucose into gluconic acid and the fructose into still more acetic acid. The bulk of the SCOBY mass is a secondary product of the fermentation process. A genetic analysis of kombucha SCOBYs revealed an overwhelming majority of bacteria belonging to a bacterial genus that converts available alcohol into acetic acid. There were also numerous *Lactobacillus* and *Acetobacter* species. By far the most prominent yeasts were *Zygosaccharomyces* (a genus that includes *Z. kombuchaensis*). Culture-based tests have also rounded up a host of organisms also found in beer making, including *Brettanomyces bruxellensis* (first isolated and labeled as a spoiling microbe from the Carlsberg brewery in the early 1900s but a common helper in lambic-style beers) and *Schizosaccharomyces pombe* (discovered in the late nineteenth century in millet beer from Africa).

• • • • •

Perhaps because their workshop is so thoroughly bathed in *Lactobacillus* species from all of the krauts, kimchis, and other vegetable ferments they are making, their kombucha has a subtlety and smoothness I haven't tasted in store-bought varieties—and have rarely encountered in even on-tap kombuchas. To be sure, their kombucha has plenty of the acetic acid organisms. But many "people treat it more as an acetic acid ferment because that is the most voracious organism in there,"

Hozven says. And simply letting the acetic acid bacteria run wild tends to push it toward the vinegary side of the spectrum. "Acetic acid is really harsh, and it's a lower pH than lactic acid," she says. "To mask it, people then just add all these really cloying, sweet fruit juices to it. And then people just drink it because they think it's good for them. And that's a bummer, you know?" she says. "Because kombucha, if you really look at it, is an organism, and its evolution is really complex. So you can coax some really fabulous flavors with them. But they're tricky. They're just tricky." For example, she says, kombuchas quickly descend from nuanced to overfermented. "They change so quickly. When one wants to get bottled, it wants to get bottled, like, that day. There's sort of a sweet spot you have to hit."

Naturally fermented kombucha can also be a different experience for consumers, Hozven notes. "One of the bummers of the modern beverage industry is that we tend to become accustomed to what it tastes like to drink carbonated beverages, like, from a CO2 canister—instead of a natural CO2 produced through a yeast breathing," she says. "And it's very different." Hozven's kombuchas go through a natural fermentation, similar to the process for making champagne. In the case of her kombucha, the fermentation is open to the environment, so most carbon dioxide that is created escapes into the air. But once the product is placed in bottles and sealed, any carbon dioxide produced is trapped. And to encourage a delicately fizzy final product, Hozven and her team reinvigorate the yeasts as the drinks are bottled.

The company's kombuchas also depart from the fruity spectrum typical of many larger kombucha labels. Three of their signature flavors are celery, beet, and fennel. And they introduce new blends and flavors throughout the year, depending on the season. When I visit, there are shelves of large glass jugs gleaming a rainbow of different hues. They have been fermenting for about a week and a half. There is basil, turmeric, thyme, parsley. "Because it's so regenerative, the culture, we will constantly take pieces out of the culture and grow them on different herbs," Hozven explains. Even if the SCOBYs in the forty large jars came from the same parent, each one contains its own ecosystem. From there, she notes, "each of those is going to make about forty bottles. You go from having one ecosystem, once they're bottled, that goes up to, like, sixteen hundred kombucha ecosystems because every single bottle will age differently in a lot of ways." That, she says, "is a lot to keep track of."

To get a sense of the care that goes into these ecosystems, we sample a finished kombucha: a lime-celery-basil blend. Hozven pours a few ounces into mismatched glasses, mine a champagne flute, which is fitting, as she uses words like *champagneness* to describe their version of kombucha. It was brewed using a fresh basil-based tea and left to ferment for two to three weeks. Then they added lime juice, celery juice, "and since neither of those juices has huge amounts of sugar in them, a shot of honey" to help feed the yeasts, Hozven says. The mixture is then bottled and left to sit for an additional week, further carbonating the liquid. It is light and gentle—neither too vinegary nor too carbonated. Not too biting or too sweet. It has a singular brightness of flavor. And it is delicious.

Some people might still recoil at the sight of brewing kombuchas, with their slimy floating mushroom-like SCOBYs. But Hozven is completely taken by them. "They're totally gorgeous," she croons, gazing at her shelves of colorful evolving ecosystems.

• • • • •

Hozven got her start working not with kombucha, but with miso. After traveling during her early twenties, she found herself drawn to the macrobiotic diet, a Japanese-created cuisine that emphasizes simple plant-based foods, along with ferments such as miso and Japanese pickled vegetables. She soon came across GEM Cultures, one of the pioneering marketers of cultures in the United States, and a new world of delicious fermented foods opened up to her. She has been experimenting her way forward ever since. "There's always new inspiration," she says. "I pickle a lot like how I cook, just with things that sound particularly good to me." When she was first figuring out the processes two decades ago, it was mostly trial and error. Even to this day, she says, "I spend a lot of the day confused. That's okay, too." One key, she says, is to "get over your fear of grossness. And have a little bit of trust in yourself." And of course in the microbes.

The food world is quick to celebrate celebrity and greatness, crowning master cheese makers and master brewers. But she is humble in the face of her microbial partners. Despite her success and renown in the ferment world, she dislikes the term *master fermenter,* she says, because "no one is a master of this. The people I know who do it the best feel like failures most of the time. And that's because they're actually paying attention—and learning."

• • • • •

The name of Hozven's company highlights the traditional expansive use of the term *pickle.* "When we say *pickle,* we're generally meaning mostly a lactic acid fermentation," she says. And that can really be applied to any fruit or vegetable. Their goal has not been to re-create specific established traditional foods but, she says, "to create, like, a nice spectrum or color and flavor, where everyone can hopefully find something that they like."

In the processing area, large food-grade plastic tubs are filled with salted, sliced cabbage that is "sweating out" and preparing to be packed for fermentation. "One of the things I really like about what we do is that most of the products don't use water—they totally reply on the moisture in the vegetables to create the brine," Hozven notes. "I think you just get a much better, more concentrated flavor. It's more work, and there's certain things that it won't work well with"—such as whole cucumber pickles, which require additional water—"but most things do," she says.

To survey their lively fermenting landscape, we enter a large, walk-in refrigerator, where the temperature hovers in the mid-60s. The shelves are stacked with tubs and jars, the floor crowded with large metal barrels resembling beer kegs. It smells subtly earthy. Amazing, Something like the microbial magic that is happening all around us. Hozven opens one jar, and we sample a pickling cauliflower, mixed with turmeric, ginger, spring onion, and mustard seeds. After four weeks, it is still light but flavorful and has a nice crunch to it. The lid is returned, and the microbes are left to do more of their work. It is a reminder of the pleasurable chore of the pickling process: always be tasting.

Not all of her team's creations are entirely experimental. They have general guidelines for some old standbys. Kraut, for example, stays in large barrels here for six to eight weeks while a succession of microbes brings the pH down to the low 3s (somewhere between orange and lemon juice). But it's not as simple as chopping and dropping. As Hozven explains, people tend to assume, "'Oh, you just stick it over there and leave it.' Yeah, that'll be *really* pretty," she counters jokingly. "Everything requires constant maintenance, for sure." They check each batch of kraut on a regular schedule. And once fermentation slows down, they keep it in a cool place and pack it as needed.

These dynamic foods are clearly meeting a growing hunger in the marketplace. But they also create a challenge in communicating about them to consumers, Hozven says. People are perpetually asking her, "How long does this stuff last?" The answer, as with so many answers in the fermentation world, is complicated. Hozven and her company generally put a "best before" date on their kraut (the only food they currently sell wholesale, which means they don't get to have personal interactions with the people buying it). "We do it at five months, which is somewhat arbitrary," she says. A jar that has been opened will have contact with oxygen, which eventually degrades the cabbage, drying it out and softening it. And it may eventually start to get some harmless wild yeasts growing on its top. But does that mean it has spoiled? Not necessarily.

"It kind of speaks to how we tend to see food as, like, 'now it's good; now it's spoiled,'" she says. "With pasteurization and expiration dates, that's kind of true: if something's living in there, then that's a problem, and so then it's spoiled. But everything we do here, that's not a helpful way of looking at things," she explains. "I sometimes like to tell people, 'You're kind of buying a process more than a product.' And so they *will* change over time. Some things will change really beautifully over time. Some things will go in a direction that might not be your thing—not to say it's a problem. You have to think about it in that way, which I think is a good way to start thinking about food, but obviously challenging. Because I get the same question a million times a day."

Similarly, she bristles at the notion that their foods are cure-alls. No matter how fiber and microbe filled. "Food in general is really nourishing," she says. "I think of what we do as a whole food group unto itself. And yes, it's absolutely part of a diet that will feed you and nourish you. But I don't like necessarily thinking of it as 'which one's best for me?' and 'which one's going to save me?' and 'how much do I need?' I think that whole mindset still feeds into our general medical model of things." Instead, she suggests, "you need to eat a really wide spectrum of food, and a really wide spectrum of fermented food as well. And maybe, like, don't sweat so much exactly what is this thing that will heal you—that's not really a holistic way of looking at food and your diet."

American Kimchi

Many other homegrown companies are also putting their own spin on traditional fermented foods from across the globe. At Ozuké, a Colorado-based company, they are always experimenting with new combinations of ingredients, blurring the line between kimchi, kraut, and a just plain mixed-up tourlou of pickled vegetables.

"It's cool to make something that has such a long-standing and international tradition," says coowner and cofounder Willow King, when I visit their production facility outside of Boulder. "It's a really expansive way to think about food when you think about all the different ways people have done it—and how successful it's been over time in so many different places." That success, she says, is the product not just of human ingenuity but also of a dynamic partnership. "All of those things have been successful as a result of interaction with human and the bacterial world. So there's something I feel like gives you confidence to experiment. Because the whole thing has been sort of one long series of experimentation."

Even though we now live in the era of refrigeration and pasteurization and globalized supply chains, and even though Ozuké products themselves are now shipped across the country, much of their work still results from the same ethos that has been driving fermenters for thousands of years: frugal preservation.

The company buys some 180,000 pounds of vegetables a year from local farmers around Boulder and the surrounding area. The bulk of this food usually comes in at the end of the season, when producers have sold everything they can but their fields are still studded with excess or ugly crops. Such a purchasing strategy, though, can lead to unpredictable yields. One year they acquired from one farmer an extra 10,000 pounds of remnants—mostly cabbage—that would have otherwise gone to waste. The idea sounded good when they finalized the deal with the farmer, and it turned out to be a sound investment. But, says Willow King, "we were sort of mortified when it kept rolling in."

What is it about cabbage that has inspired so many fermentation traditions around the world? "Cabbage is a wonderful thing," says cofounder Mara King (no relation). For example, the outer leaves, even if they go soft, can be pulled away to reveal fresh inner layers. A sort of built-in wrapper to keep the inside crisp for weeks and even months. After receiving the bumper crop of cabbage, for example, the Ozuké team left the thousands of heads of it in the cooler from November through January, when they finally had time to process it. And over the two or three months, the cabbage heads lost only about 15 percent of their weight (through discarded leaves and water loss). "It holds up," Willow King says. "I thought we would have trimmed a lot more," adds Mara King—the business partners and longtime friends are quick to finish each other's thoughts. The company actually came not out of a market analysis or sector assessment, but out of an informal group they had with their kids. The two moms would get together with their young children and undertake different culinary projects. And fermentation quickly became the favorite process. "They were fun because they were relatively easy to make, and they were exciting because they did this weird, bubbly food science," Willow King says. "Our families were into them. We were into them. They were really tasty. They made us feel really good . . . And we started just tinkering around."

Thanks to their ability to stockpile and then use the food—even if a little bit was lost along the way—they averted the alternative, which would have been 100 percent of those thousands of pounds of cabbage being wasted. This is a common theme in the current U.S. food system, where only some 50 percent of food that is grown makes it to your plate. From there, another substantial percentage gets tossed as well. "I think fermentation is one way for people to really gain some ground and preserve," says Willow King. "It's less labor- intensive than canning—and less nerve-racking because there's less that can go wrong. It's an awesome way for people to get that food empowerment—whether they garden or they just want to buy stuff that's cheap produce at the end of the season and make their own."

• • • • •

At Ozuké, kimchi is the best-selling pickle. To make it, they usually ferment their chosen veggie blend for a week or two, depending on how warm it is and what ingredients are in the ferment. The process pretty much takes care of itself. "If it has everything it needs, it just goes for it," Willow King says.

But the zeal of the microbes can be strong—sometimes a little too strong. So the cultivating humans have to be careful to ferment the product long enough before packaging it in jars and shipping it out. With fermented beets, especially. The Kings and their early customers learned the hard way that the beets need to ferment a lot longer than cabbage (for four to six weeks—and then again in a slower, secondary fermentation in the cooler for another four to five weeks.) "There's just so much sugar in there that if you don't let it ferment . . ." says Willow King. "If you under-ferment it," Mara King jumps in, "you create jar bombs." That's because if the fermentation hasn't slowed before packaging, Willow King explains, "it keeps fermenting in the jar," with the microbes producing more and more carbon dioxide, building pressure in the sealed container. "We've definitely heard from people in the early days, well, and still occasionally, someone's like 'Thank you for painting my entire living room purple—and the entire outfit I was wearing.' Sorry! 'It's fine, we won't charge you extra for that!' " she jokes. This also helped encourage their move from a shared commercial kitchen space to their own facility. They occasionally filled a beet fermentation vessel too high, and "if you walked in the next day," Mara King says, "it literally looked like a murder scene."

Although they have gotten a better handle on the macro dynamics of their ferments, Mara and Willow are eager to learn more about the microworld contained in their jars. They have been able to do basic colony counts from their various products. They found, for example, that their sauerkraut with juniper berries has the lowest microbe count—at about 500,000 colony-forming units per gram of kraut (CFUs are the count of how many organisms are live and vivacious enough to reproduce). This isn't surprising, Mara King says, as juniper berries are known to contain antimicrobial compounds. On the other end of the spectrum, kimchi was the most microbially robust, with some 88 *million* live microbes per gram. "It always ferments faster and has a higher *Lactobacillus* count than any of the other ferments that we do," she says. They suspect this is because the kimchi especially contains so much good food for the microbes.

They are eager to learn more about the microbes that help create their products, but they have yet to be able to invest in strain-specific scans. On the other hand, though, "in some ways we're approaching it all backwards," Willow King says. "People have been doing this from the beginning of time, right?" Or at least long, long before genetic sequencing and microscopes. Mara King agrees. "We're using an ancient technique." And the microbes' work hasn't changed.

Wild Brews

Today there are also lots of new-wave fermenters reaching back to microbial traditions not in foods or health tonics, but in good old-fashioned beers. And by old-fashioned, I mean really old-fashioned.

Before the discovery and isolation of brewer's yeast, *Saccharomyces cerevisiae*, beer was something to be guided—not locked down in sterilized metal tanks and temperature-controlled distribution channels. And some brewers are rediscovering the wilder tradition.

Although lambic beers are perhaps the most famous of the wild fermented beers, they come in many more stripes and flavors. On a stretch of redeveloping businesses in Denver is a storefront turned taproom. Here in a side room, the microbes run the show.

Coaxing the microbes to do his bidding is James Howat, a bearded, beanie-wearing former high school science teacher. Howat studied microbiology as an undergraduate and was an experimental home brewer for years. When he and his wife originally opened the brewery and taproom, they offered familiar standards such as IPAs and saisons. Soon, though, he was experimenting again. This time not with known recipes and store-bought strains, but with the microbes around him. The products were truly wild-fermented beers.

He began pursuing these wild-fermented brews as a side interest, which he and his wife Sarah dubbed the Black Project, for its secret status. But it didn't stay under wraps for long. In their early limited releases, these brews gained popularity, traction, and soon recognition at the Great American Beer Festival, held just up the road in downtown Denver. So, two years after opening their brewery, the couple bet the microbial farm on wild ferments. And it has been a success.

To make his beers, Howat relies on a little bit of history, a little bit of science, and a lot of location. As is true with miso makers in Japan—and countless other fermenters around the world—Howat's beers have their own terroir, specific to the ambient microbial communities.

He begins making his beer much the same as any beer brewer, by combining the grains, hops, and heated water to make the wort. But soon the process deviates. Instead of going right into a closed fermentation vessel with specific strains of yeasts, Howat's wort gets transferred to a coolship. This large, open copper fermentation vessel leaves an expansive surface area of the liquid exposed to the air, allowing it to gather environmental microbes as it cools. Historically, these open-air inoculators were often located in old wooden attics, where a local microbiome would settle in. The Howats' location wasn't graced with a four-hundred-year-old attic (or any attic at all), so when they were just starting out, they would cool the beers on the roof of the building overnight, covered by a thin layer of mesh, allowing the neighborhood's microbes to find their way into the liquid.[‡] Now, with a 300-gallon coolship, the brewery designated a corner of the aging room as its home. Its contents cool under an open window—and under a freshly installed wooden ceiling, designed to grow its own microbe community to begin to mimic the old wooden attics of old.

Once cooled, the inoculated liquid goes into steamed wooden wine barrels (steamed to remove the bulk of the wine microbes, allowing more of those acquired by the beer to take hold) in the aging room. Although the precise microbes and dynamics are always slightly different, the fermentation does follow a generally predictable process, Howat explains. In the first few hours, the bacteria that are ascendant make compounds that, he admits, "are not great tasting—or smelling." But, he says, "eight months later, [the yeast] *Brettanomyces*[§] can convert those into something that's really unique. You can't get that unless you have those precursors which you wouldn't want to put in a beer otherwise. You can't do it intentionally, really. You just have to let the ecosystem do its thing."

And that's not an overstatement. "Part of what I'm learning about—and learning our cultures—is how I can start to manipulate slightly the evolution of what's going on in the tank," Howat says. He does this with some of their non–purely spontaneous beers by a process known as the solera method, which is sort of like the inverse of backslopping. Instead of leaving a little beer behind, he takes just a little out and then adds in

‡ There are plenty of stories about what found its way into old-fashioned lambics in the cobwebbed, pigeon-filled attics of yore.

§ Often just referred to by brewers as Brett—like another buddy helping out in the brewery—this genus is considered a spoilage yeast in most beers but is prevalent in lambics and other wild ales.

new coolship wort. Keeping a critical mass of the aging beer, he says, helps to keep the consistency. It also allows him to fine-tune the flavors to a certain extent. If the brew starts to become too acidic, for example, he can create a wort that is a little bit less digestible, favoring a yeast fermentation over a bacterial fermentation, which tamps down the acidity. His beers veer toward *Pediococcus* species (also found in sauerkraut) rather than *Lactobacillus*, which would lend an even sourer flavor.

Howat admits that he has put himself in a funny position. "It's kind of weird being a microbiologist, who, knowing a lot about the clean fermentation side and understanding single strains, to be doing this stuff with the cultured beers. We don't really know what's in there at all." Having previously been in full control of his standard beers, he sees his new role in high contrast. "I think of making clean beers and IPAs is like being a rancher of yeast, where you have one species. I think that coolship spontaneous ales are more like being a rain-forest ecologist. The thing that I find so fascinating and so inspiring about those beers is it's truly like an interaction between all these different genera and species and strains that isn't even really fully understood."

But do his rain forests ever get a little *too* wild? Sure, he says. "Sometimes it doesn't do its thing very well, and that's part of the risk of making these beers." But, he says, about 90 percent of the time their brews end up going down the hatch, rather than down the drain.

• • • • •

So what does the microbial terroir of Denver taste like? "There's definitely a certain peach or apricot character that I get in a lot of our beers that I don't get in other beers," Howat says. "There's just probably some strain that's growing on some tree or something nearby that keeps ending up getting in our wort," he says.

"That's the real exciting part to me is these couldn't be replicated elsewhere. I could tell somebody how to make it, and they could follow our process exactly, but geographically, it's different. It wouldn't be the same, which is kind of neat." The difference isn't always obvious, he says. "But it's something in there. It's almost a ghost of a thing."

Whatever ghosts he is collecting from the environment are hard workers, creating interesting and complex beers. Beers that also happen to be swimming with a dynamic ecosystem of live microbes.

In the Kitchen

On a chilly October afternoon as the sun starts slanting downward, local foodies and fermenters begin to assemble on a small farm outside of Boulder, Colorado. A couple of turkeys strut in their pen nearby as volunteers set up tables and sign-in sheets. It is the culminating event for Cultured Colorado, a weeklong fermented food festival in cities along the state's Front Range. Community members arrive to mingle with producers, ask very personal questions about kahm, and to meet the apostle of the modern-day fermentation movement, Sandor Katz.

As far as apostles go, Katz is a supremely humble and down-to-earth one, sporting his signature muttonchops, and, that evening, he wears a busy button-down shirt and a blazer adorned with painted leaves and fabric folk art figures stitched on. He gracefully ducks away from a growing crowd of fans to chat with me at a folding plastic table under the trees.

Katz traces his love of fermented foods back to eating kosher dill pickles as a kid in New York City.¶ But pickles are not where Katz started

¶ This style of pickling was brought over by Eastern European and Jewish immigrants. The New York

his own fermentation odyssey. His journey began one day when he found himself with a bumper crop of cabbage.

• • • • •

After being diagnosed with HIV in the 1990s, Katz left his career and his life in New York City to move to a commune in rural Tennessee to focus on his health. His idea was cleaner eating, lower stress, and some fresh air. Fermented foods were not part of this original picture.

Then in the midst of his first year's harvest in Tennessee, he looked across his garden one day and realized the cabbages were ready to be picked. All of them.

There was no feasible way he was going to be able to eat all of them fresh. Or even not so fresh. So he thought to himself, *I guess I should learn how to make sauerkraut* . . . And he did what any baby boomer would do when looking for guidance in the kitchen. He turned to the *Joy of Cooking.*

The classic, convivial kitchen chaperones of the twentieth century, Irma Rombauer and Marion Rombauer Becker, recommended he "slice the cabbage finely into 1/16-inch shreds and mix with salt. Pack firmly into stone crocks." They note that "the best quality kraut is made at a temperature below 60 degrees and requires at least a month of fermentation. It may be cured in less time at higher temperatures, but the kraut will not be so good."

So he followed their instruction, and after his first batch of kraut, he hasn't looked back. "I'm really devoted to fermenting vegetables," he

---

kosher dill was historically sold in two phases of fermentation—early (the half sour), leaving a bright green color and a crisp texture, and late (the full sour), producing a softer, more tangy mature pickle.

says. His nickname could probably also tell you as much: Sandor Kraut.

• • • • •

Since his first fateful batch of sauerkraut, he has experimented with countless other foods, taught classes across the country, and written books, including *Wild Fermentation* and *The Art of Fermentation*. As he writes in *Wild Fermentation*: "Sometimes I feel like a mad scientist, tending to as many as a dozen different bubbly fermentation experiments at once." The lesson from all of his assays is that just about any vegetable or fruit packed into a salted liquid can—and will—be pickled.

Another hallmark of the fermentation process—beyond preserving—that Katz highlights is that it can also render as edible parts of plants that otherwise might go to waste. In *The Art of Fermentation*, he extols pickled watermelon rind as a legitimate rival to classic cucumber pickles. Similarly, tough stalks yield to a luscious sour crunch when transformed by pickling microbes.

For all of his devotion, though, Katz doesn't tuck into a big bowl of sauerkraut or kimchi for breakfast, lunch, and dinner. Like people have done for thousands of years, "I use those as a condiment," he says. "If I'm eating a sandwich, or I'm having my eggs or whatever I'm eating, I'll put a little bit of fermented vegetables in with that. Maybe a little bit of miso. I sort of mix and match."

And moderation is especially key when dealing with some of the more potent flavors of fermentation. "There certainly have been some that I've had to challenge myself to try because the smell is so strong," says Katz. "I think of surströmming," the fermented herring sold in bulging cans in Sweden. "But that's great. I've really enjoyed it. It's a great complex flavor. I'm not sure it's something I'd want to eat a giant portion of, but that's not the way the Swedish people eat it either."

• • • • •

Since his immersion into the world of fermented foods, Katz has enjoyed fairly good health. But he is quick to note that fermented foods are no silver bullet. "No, fermented foods have not made HIV go away," he says. He takes regular medication for his infection and has maintained a balanced attitude toward his diet. As he explains it, "We know that the bacteria in our guts, in our intestines, have a lot of influence over different processes in our bodies and that they're related to digestion, to nutrient assimilation, to immune function, to mental health, to liver function. There's some evidence that foods that can positively influence the community of bacteria in the intestines might help in any of those areas." So, he says, "if you approach someone with the idea that 'oh, okay, maybe this can help my overall digestion and nutrient function; maybe this can help my overall immune function; maybe this can help my overall mental health,' then that's huge. Whatever the status of your health, those things are of great benefit—and there's really no appreciable risk involved."

After all, he says, for thousands of years "everyone ate 'bacterially,' just because [bacteria] were part of how people were preserving foods for food safety, making foods digestible, and making foods delicious."

Despite all of the science that is now allowing us to peer into these busy ferments, "it's an ancient way of thinking," he says. "A lot of cultured foods have been associated with good health and longevity—and there's lots of reasons why that could be so, whether it's kombucha, whether it's sauerkraut, whether it's yogurt, whether it's kefir, whether it's pulque in Mexico."** And now more than ever, he says, it is crucial to introduce these foods back into our repertoires.

** Pulque is a drink made by fermenting the sap of the agave plant; it contains Lactobacillus acidophilus, Lactobacillus mesenteroides, Lactococcus lactis, and other microbes.

• KRAUT-CHI •

Sandor Katz takes a broad view of making pickled products. His fermented cabbage variation is a hybrid product he calls "kraut-chi"—a blend of sauerkraut and kimchi—and is an adaptation of what he considers his most basic recipe for pickling. You can follow the basic sauerkraut process but get creative with additional ingredients, whether that is sliced beets, chili powder, or fermented fish.

To make his kraut-chi, Katz follows the same basic process he prescribes for making sauerkraut in *The Art of Fermentation*.

Chop vegetables.

Add salt and massage until enough liquid has emerged (or, skip the massaging and just add a little brine).

Pack the contents into a container, and wait until the flavor is one that you like.

It's not often that a recipe can be boiled down to four words, as he concludes: "Chop, salt, pack, wait." But, as he adds, don't forget to "taste frequently, and enjoy!"

Above all, be flexible, Katz admonishes. "I do it a little bit differently every time," he says. And even if you followed the same ingredient list and procedure precisely, the results will never be exactly the same, owing to differences in the ingredients, their microbial residents, and slight variables in the environment. So relish the variety in this dynamic, collaborative processes.

No matter what recipe (or non-recipe) you follow, Katz says, "It's really what I would recommend to anybody who's interested in trying their hand at fermentation—just because it's incredibly easy. You don't need special equipment, you don't need special starter cultures. It's incredibly delicious. It's rich with diverse bacteria."

To Ferment or Not to Ferment?

With all of the inventive and delicious fermented foods cropping up at farmer's markets, health food stores, and even mainstream supermarkets, it might seem an unnecessary step to try fermenting at home. And perhaps a little bit daunting.

Cast aside your concerns. Drop your doubts into the compost bin. So many of these flavorful and fascinating foods are incredibly easy—and satisfying—to create yourself. And will yield far richer results. As chef Nobuaki Fushiki reminds us, we don't have do it all ourselves. We have many (microscopic) cooks working with us.

• • • • •

Many a first-time fermenter will surely have some questions about safety. Most all of us have been brought up to refrigerate everything, check expiration dates, and consider toxic anything that has the slightest blush of mold. So how can leaving food out in ambient temperatures for days, weeks, or even months at a time produce something that is safe, never mind nutritious?

Well, it is time to push our spectacles back up our noses and revisit some of the microbiology we have learned. Namely, acid kills pathogens. So as the acidity level in a ferment rises—whether in yogurt or sauerkraut or kombucha—pathogenic microbes bite the dust. This, in fact, is one reason that fermenting is considered safer than canning. If foods are canned improperly, *Clostridium botulinum*, source of the deadly botulism toxin, can survive. In fact the U.S. Centers for Disease Control and Prevention recommends any canned food with a pH above 4.6 be pasteurized with a pressure canner (even the boiling water method is out these days). Ferment your veggies to a pH lower than that, and you're in the clear. So with the specter of botulism vanquished from the landscape, we are free to proceed without fear of food poisoning or death. As Sandor Katz notes, "It's extremely safe. There's no case history of illness or food poisoning" from properly fermented vegetables. So as long as the acidity levels reach a safely low pH (many authorities recommend 4.6 or below), your ferments should be in the clear.

"The other great advantage of fermented foods," Katz says, "is that they've been predigested by bacteria before we eat them. In the crock or in the cheese or in the salami or whatever is happening, nutrients are being broken down into simpler, generally more bioavailable forms." As with lactose-digesting microbes in fermented dairy products, microbes in other fermented foods, such as cruciferous cauliflower and cabbage, can make them easier for many people to digest.

Beyond these practical reasons, Katz asserts that there is another, less tangible reason to make and to eat fermented foods: they are part of our human cultural heritage. As he writes in his book *The Art of Fermentation*, "I have searched—without success—for examples of cultures that do not incorporate any form of fermentation." These foods "appear to be found in some form in every culinary tradition." So why should we abandon the microbes now?

Feeding the Starving Gut

All of the fermentation fervor has no doubt been a boon to the transient microbe populations in guts around the world. And probiotics are such an exciting, mysterious, and dirty world to explore.

But this burst in popularity has not extended so much to prebiotics, leaving these essential components largely on the sidelines. It is perhaps not so easy to market whole grains and raw onions with the same élan as a punchy pomegranate kombucha. But that's a shame, because by overlooking prebiotic foods, we are ignoring our full-time gut inhabitants, who are trying desperately to eke out a living. And sure, maybe it's tough to market a knobby tuber—or a sprinkle of unsweetened cocoa on an otherwise unremarkable bowl of oatmeal. But some are trying.

One team of chefs has been bringing a refined view of fiber and culture to a discerning clientele.

In a smartly decorated restaurant along San Francisco's once-maligned, now-gentrified Mission Street, the chefs at Bar Tartine served up a thoughtful and sophisticated menu—with a side of consideration for the gut and its denizens.

Chef Cortney Burns discovered this way of approaching food while dealing with gut issues of her own. After eating with purpose for her lower intestine and its residents, she says, "for one, I felt great. Then I stared realizing the vast diversity of things that you could do with it from a culinary standpoint. Not only was it good for you, it ended up also being great for harnessing aroma, flavor, texture—all of these things that you couldn't do any other way. It just became a passion."

On one spring evening, their Friends + Family tasting menu offered up fourteen different dishes, all perfectly parceled out—even for a party of one squeezed in at their long white marble bar. Similar to a Japanese meal, dishes arrived in clusters of small bowls. A beet soup with caraway seeds and fresh buttermilk was tart yet creamy. An array of whole-grain breads offered a range of textures and fibers. Red, house-fermented sauerkraut with ginger was bright and fresh tasting. A chicory salad with anchovy and rye was topped with a tangy feta. And a cultured sunchoke dip left a hint of artichoke on the tongue, its silky smoothness undercut with a subtle bite.

Cooking culturally, with our microbes in mind, is not something that comes as first nature to many people raised in the United States. It involves relinquishing a certain amount of control. Offering some of the process up to organisms we cannot see—forces that seem mysterious and possibly even slightly dangerous. And weaving in foods that also contain prebiotic compounds can take even more getting used to (and not just from a textural perspective). It is easy to make a dish taste good with rich, refined ingredients. But to bring in chicory or sunchoke or a big serving of legumes requires not just an intentional hand in the kitchen, but also a trust in people's willingness to eat a little outside of the expected. Thankfully, Burns and her colleagues took that chance—and if the weeknight crowd at the restaurant was any measure, it paid off for all.

Although Burns's dishes were exquisite and highly accomplished, there is no reason we cannot take some lessons from them to apply in our own daily lives. That can mean broadening the palette of plants in your repertoire—whether that's stocking more dried legumes, buying (and eating) those greener bananas, or trying more obscure vegetables, like sunchokes.

Some of those ingredients don't look terribly appealing at first glance. "They are not really friendly when you look at the roots," says microbiologist Patrice Cani when I meet him at the Keystone Symposia on the human microbiome. "Not really friendly, but tasty, really good."

Sunchokes are one of these unfriendly looking root vegetables. These knobby little tubers are a terrific source of prebiotic fibers, particularly inulin, as well as fructooligosaccharides and a wide range of vitamins and minerals.

And as I discovered, they are exceptionally easy to grow. A little too easy, in fact. On a quest to experiment with a more microbe-friendly diet, I cast off admonitions to not plant these sunflower relatives if you didn't want your garden forever overtaken by these tall, hardy plants. I ordered a small packet of starter tubers and planted them in the late spring next to my potatoes, unsure as to whether they would take or get crowded out. But take they did, towering up and over our highest, six-foot trellis, eventually producing disappointingly small yellow flowers just past the peak of summer. That first fall, I harvested a basket full of the roots. I mistook their subterranean vigor for overall sturdiness and left them in a back room, still soil-caked, for a week or so before I could get around to figuring out exactly what to do with these pounds upon pounds of inulin-containing roots. When I returned to them, I found a basket of shriveled and unusable brown lumps. So into the compost bin they went. Perhaps that was it for the sunchoke experiment. The following spring, however, it was immediately clear that I certainly hadn't harvested all of the roots because the plants started filling in once again, spreading now to challenge the potatoes. Then the turnips, then the beets, and then the parsnips. Come summer, the bed was a mass of ungainly stalks topped with uncutworthy flowers, blocking most all of the light to my nearby basil and rosemary. It was then that I finally understood the name *sunchoke*.

I vowed all summer, cringing at the eyesore of our front-yard sunchoke patch, to take drastic measures come fall. I would even dump the carefully cultivated soil bed if necessary. Anything to liberate my garden from this insidious plant.

But come November, in a fury of pre-snow gardening and processing, I harvested yet another basket of the tubers (this time far more numerous than the last), and got them washed and scrubbed down right away. I stared down at two large mixing bowls full of still-wet golden, gnarly roots, wondering what on earth to do with them. I confirmed on homesteading websites that no, they do not keep well aboveground. (Apparently the best thing to do is to harvest them as needed through the fall and winter—or if you were a novice like me and picked a boatload at once, you could bury them in a tub of sand and keep them on your porch.) So I took a chance and cooked them all. And was I glad I did. A bite through the skin revealed a soft, sweet center that was much better than a potato. An experimental batch of large, whole-roasted sunchokes went straight into the fridge to serve as a side later that week.

And a full tray of smaller sunchokes got packed into snack bags for lunches and into larger freezer bags for dinners throughout the winter.

So the sunchokes earned the three-by-ten-foot bed they were rapidly commandeering. And the following summer, my cringes were tempered slightly thinking of the unexpectedly delicious prebiotic snacks we would be enjoying again soon enough.

ROASTED SUNCHOKE

Here is an easy preparation for roasted sunchokes to serve as a side or to toss into a salad. You will need sunchokes (also known as Jerusalem artichokes), olive oil, your choice of herbs (rosemary or thyme work nicely), and sea salt and pepper to taste.

Preheat the oven to 350 degrees F.

Wash about one pound of sunchokes, and remove any eyes as you would from potatoes.

If sunchokes are large, chop them into smaller pieces (roughly one inch works well)

Whisk olive oil and herbs together in a large bowl, add the sunchokes, and toss to coat.

Spread the sunchokes on a rimmed baking sheet and sprinkle with sea salt and pepper if desired.

Roast for about 30 minutes or until tender.

Although eating for our microbiome might have once been built into our cultures, we are now faced with more food choices than ever, many unrooted from their finely tuned traditions. This can make it challenging to find a mooring and reassemble a well-rounded diet rich in fermented foods and prebiotic ones. But it helps to know that we are not just what we eat, but also what our microbes eat.

CONCLUSION

Saving an Invisible World

We are quick to decry the loss of biological diversity in the world's rain forests, coral reefs, and Arctic ice floes, recognizing that these global shifts—even in faraway places—have long-term implications for our own well-being. But we have been slow to wake up to another rapid extinguishing of life that is occurring right in our very own bodies, one that appears to be having an immediate and very real impact on our physical and psychological health.

We have unwittingly transformed our microbial profile. We are often delivered by C-section, are fed on formula (or breast milk that no longer contains the full complement of bacteria), are administered antibiotics for each sore throat from infancy through old age, live and work in pristinely clean environments, douse ourselves in antimicrobial products, eat food that has been processed, refined, pasteurized, pressure-treated, and irradiated. And we consider ourselves marvels of modern science as we push the average life span up past seventy-eight.

At the same time, obesity and metabolic syndrome have skyrocketed, depression is pervasive, kids are getting type 2 diabetes before they learn the multiplication tables, and EpiPens are commonplace in classrooms.

Perhaps we're not doing as well as we think we are. We have innumerable things to blame for these conditions, from sitting too much to being exposed to poorly studied chemicals. But many researchers are looking to the microbiome as a mediator in many things that seem to be going wrong with our health. Although helping your microbes isn't going to steel you against every disease, gained pound, or blue spell, surveying how we are doing as a society, Erica Sonnenburg says, "I don't think you can make the argument that the Western microbiota's just fine."

One of the main problems is that we are losing diversity, both in our resident microbes and in the microbes we encounter through food and through the environment. As Erica and Justin Sonnenburg note in their writings, "as diversity is lost in the Western microbiota, this ecosystem is at greater risk of collapse."

• • • • •

Although the microbiome remains in many ways a black box, we are learning more every day about our beneficial bugs and their relationship to our health. And some interesting information has surfaced. For example, lower diversity is correlated with obesity and poorer health as well as certain profiles in IBD. Still, there are no clear-cut prescriptions for a healthy microbiome, and it's likely that a healthy microbial community will look a little different for everyone. And maybe someday we will be able to dial our microbes in even more precisely, identifying which are the most helpful for each of us.

One place we can already start is in getting to know our current communities. For less than a hundred dollars, anyone can now have their gut microbiota sequenced, learning about their most prevalent species, most unusual species, and overall diversity compared with that of other people around the globe. This is, of course, just a static snapshot. A round of antibiotics, a trip abroad, a few days off your normal diet, an illness, or a bout of stress can change this profile. Someday you might have daily readouts that monitor microbial health from the privacy of your bathroom. This longitudinal information could enable us to track trends, perhaps detecting changes that alert us to impending health issues—or help us fine-tune our diet for personalized optimal gut microbe profiles.

In addition to monitoring our own microbes, we will continue to discover new microbes for use as probiotics. Beyond that, we might soon even be looking at microbes that have been tweaked in some way to provide additional benefits. Some researchers are already starting to genetically engineer strains, outfitting them with helpful traits. Scientists will also find more targeted ways to manipulate the microbiota as a whole, through diet and through specific prebiotic compounds.

• • • • •

We are in the midst of many unplanned global-scale health experiments. Some are heartening, such as the eradication of once-common infectious diseases. Others are horrifying, such as the spread of ailments related to obesity—once an exceedingly rare condition that now affects 13 percent of the global population (and 35 percent of the U.S. population). Still other health experiments that we didn't sign up for have also been going on under our noses, quietly perturbing ancient systems, creating outcomes we are just starting to piece together.

There are no hard-and-fast prescriptions for rescuing a stumbling microbiota or burnishing a decent one. But looking around the world, into the many cuisines that human cultures have created, we can take some inspiration from these generations upon generations of informal, evolution-driven experimentation.

Research into diet and the microbiota backs up many of the wisdoms common to traditional longevity- and health-linked diets: Eat lots of fruits and vegetables, get plenty of fiber, take meat in moderation, and augment your meals with something fermented. “We’re all connected,” says Erica Sonnenburg of our microbes and us. “And so something that isn’t healthy for the human body is likely going to be bad for the microbiota.” And vice versa.

For our ancestors, choosing kraut over cake or kimchi over chips wasn’t much of an issue. Our food options were limited to what we or our neighbors could make or collect. Today we’re up against a new set of challenges. “You go to a bakery, and it looks good to all of us, right?” she says. If we evolved to seek out sweet foods for immediate energy and fatty foods for bodily reserves—in part because these things were so scarce—how can we now make better daily decisions? A lot of it comes from practice and early education, she says. “That needs to be instilled in the children, so that when they are adults, they say, ‘I’m going to choose what I know is going to make me healthy, not what the ancestral part of my brain craves.’ ” Children in many cultures are taught to value health over instant gratification. “That’s definitely the tactic we use with our kids,” she says. “When they’re faced with a kale salad, we talk about it going to feed your microbes. You’re going to be healthier eating these things, and you’re doing this for the greater good—not for the immediate reward of sugar-and-fat endorphin rush that we all have.” Justin Sonnenburg chimes in: “We compare it to indoctrination for religion. It’s got to be *total* brainwashing,” he says with a wry smile. “It just becomes such a fundamental part of who these people are that when they are faced with a choice, they say, ‘Oh, no, I’m a salad person. That’s what my people do.’ ” It becomes *their* culture.

Eating for our microbes doesn’t need to be arduous or devoid of pleasure, though. Once you know what to look for, it's easy and can be quite delicious. Sauerkraut, kimchi, and other fermented vegetables add another dimension of flavor to a meal. Fermented bean pastes bring a new richness and umami to otherwise drab dishes. And fibrous vegetables lend a complexity and hardiness to recipes. Some might find the flavors and textures take some getting used to. And to be sure, they are not intended as a main course. But once you become accustomed to that fermented or fibrous bite, you might start to find meals seem lacking without it.

• • • • •

It is important to not lose sight, amidst all of the seething new fermented food trends, of these dishes' places within the context of their cultures. Most people in Japan would surely cringe at the idea of an all-natto dinner, and a breakfast of kimchi alone probably wouldn't cut it in Korea. These foods are a pervasive and consistent part of these cuisines. But they are just a part. They take their place alongside a diverse offering of foods, which in turn provide a diverse range of fibers and macronutrients. We would do well to eat broadly and not relegate our fermented or fibrous foods to the occasional health-kick meal. We are what we eat, so let's eat diversely.

• • • • •

Just as we are losing macro species around the planet, we are also experiencing a great die-off of microbial species that have been with us, helping us and our ancestors, for millions of years. In the midst of this cataclysm, scientists are working furiously to catalogue the microbial communities of people across the globe, from traditional hunter-gatherers in Tanzania to Amazonian tribes in Venezuela, from you to me, before these species and strains vanish forever.

Beyond reducing diversity in our own guts, globalization and industrialization are also threatening the wide range of microbial food cultures. These are the microbes that have for thousands of years helped to maintain traditional dishes and processes. As companies find ways to streamline production with rigidly defined strains, and as regulatory bodies insist on controlled, sanitized procedures, the complex traditional amalgams of robust microbes are vanishing from the food landscape, too.

As we've seen, many scientists are diving into the microbiology of traditionally cultured foods and beverages. But even today, this exploration is still often done with an eye to create standardized starter cultures. Simplifying these varied products is a move in the wrong direction. We are just starting to learn how diverse and varied these cultures truly are—and how essential diversity is to our food and our health. When we lose these microbes, we lose all of the subtle flavors, textures, and experiences of food only they can create. And their presence in our bodies.

What else do we lose when we lose the microbes we have lived with for so long? We lose many of these time-honed expressions of ingenuity, of the broader human culture.

We might be losing a little bit of what has made us human.

So it is a shame to barge in with abbreviated uniform starter cultures that homogenize all matzoon, natto, and chicha, reducing our exposure to a wide array of microbes, while possibly stomping out rare and as-yet-unidentified strains and interactions—and health benefits. Such a cultural imposition would hardly be tolerated today on a macro human or environmental scale. We need to pay more heed to the shifting landscapes of microbial diversity and tradition. With it, we will be preserving a richer future for us all.

So, with nothing to lose and billions, trillions, of microbes to gain, let's get (re)culturing.

FURTHER READING

Below are some of the popular press books I consulted in my research that I found most valuable. I have kept this list to those more focused on the science and the food rather than on diet books (some of which draw questionable conclusions based on the scientific data currently available). As interest in this field grows, more excellent books about the gut microbiota, fermentation, diet, and food come out every season. So I encourage you to keep reading—as you would eat—for your microbiome: diversely.

Martin J. Blaser, *Missing Microbes: How the Overuse of Antibiotics Is Fueling Our Modern Plagues*. New York: Henry Holt, 2014.

Dan Buettner, *The Blue Zones: Lessons for Living Longer from the People Who've Lived the Longest*. Washington, D.C.: National Geographic, 2008.

Rob Dunn, *The Wild Life of Our Bodies: Predators, Parasites, and Partners that Shape Who We Are Today*. New York: HarperCollins, 2011.

Gulia Enders, *Gut: The Inside Story of Our Body's Most Underrated Organ*. Vancouver: Graystone Books, 2015.

Masayuki Ishikawa, *Moyasimon: Tales of Agriculture*. New York: Del Rey, 2009.

Sandor Ellix Katz, *The Art of Fermentation: An In-Depth Exploration of Essential*

Concepts and Processes from Around the World. White River Junction, VT: Chelsea Green, 2012.

Sandor Ellix Katz, *Wild Fermentation: The Flavor, Nutrition, and Craft of Live-Culture Foods.* White River Junction, VT: Chelsea Green, 2003.

Rob Knight, with Brendan Buhler. *Follow Your Gut: The Enormous Impact of Tiny Microbes.* New York: Simon & Schuster/TED, 2015.

Daphne Miller, *The Jungle Effect: A Doctor Discovers the Healthiest Diets from Around the World—Why They Work and How to Bring Them Home.* New York: William Morrow, 2008.

Justin Sonnenburg and Erica Sonnenburg, *The Good Gut: Taking Control of Your Weight, Your Mood, and Your Long-Term Health.* New York: Penguin Press, 2015.

ACKNOWLEDGMENTS

First, to the many, many microbes responsible for my being here to write this book, thank you. I am very sorry for the questionable "sick" claims I made as a kid that resulted in questionably necessary antibiotics. And gratitude to the multitudes who have contributed to all of the delicious fermented foods and drinks I enjoyed while researching this book.

Beyond the biological, this book would not have been possible without the incredible generosity (and often brilliant insights) of the dozens of experts (in science, food, or both) who spared a little bit of their time and knowledge to show me their corners of the microbe universe. Thank you to Ueli von Ah, Katherine Amato, Elizabeth Andoh, Martin Blaser, Cortney Burns, Patrice Cani, Jonathan Eisen, Nobuaki Fushiki, Bruce German, Daniel Gray, Matt Hann, Colin Hill, Kenya Honda, James Howat, Robert Hutkins, Alex Hozven, Purna Kashyap, Sandor Katz, Mara King, Willow King, Rob Knight, Cristophe Lacroix, Myung-Ki Lee, Eric Martens, François-Pierre Martin, Masu Masuyama, Takahiro Matsuki, Daniel McDonald, Annick Mercenier, Leo Miele, David Mills, Eliane Murith, Kostas Papadimitriou, Enea Rezzonico, Yukari Sakamoto, Hideyuki Shibata, Erica Sonnenburg, Justin Sonnenburg, Daniel

Stalder, Kelly Swanson, Effie Tsakalidou, Peter Turnbaugh, Nicos Vallis, Jens Walter, Benjamin Wolfe, Tòmas de Wouters, Gary Wu, Amir Zarrinpar, Liping Zhao, and everyone else who contributed to the journey. Even those many amazing conversations that did not make it into the pages of this book helped immensely in shaping my understanding of this complex and evolving topic.

Many thank-yous to my amazing agent, Meg Thompson, for finding me and for finding homes for my books—and for lending a patient ear when needed. Thank you to Pam Krauss for acquiring this book and starting it on its lengthy fermentation process. And many, many thanks to Marian Lizzi for seeing this book through to its final finished product—and making sure it was ready for consumption.

I am grateful to each of the news editors I have worked with on microbiome stories since I began covering the topic as an intern at *Scientific American* in 2009, including, among others, Scott Hensley, Robin Lloyd, Ivan Oransky, and Phil Yam (whose tweet I still keep pinned up in my home office: "Trending on @SciAm: Artificial gut explains dark-chocolate benefits. 1st time I let 'poo' stay in copy . . . by @KHCourage"). Thank you, too, to colleagues past and present who have encouraged and supported me along the way (particularly to those at my day job at the College of Natural Sciences at Colorado State University). And a deep bow of gratitude to all of my incredible teachers and advisors over the years, including Mark Amodio, Frank Bergon, Mark Johnson, Derek Perry, Peter Robinson, Steve Weinberg, and many others.

Thank you, thank you, thank you to my wonderful parents, Pamela Rogers and William Harmon, who raised us kids to be curious—and to not be afraid of getting dirty. Unending gratitude to my grandparents Theodore and Elizabeth Rogers for your enduring support—and for each being an absolutely endless source of inspiration for a long life of learning, passion, and deep appreciation (and of course to keep running and to keep writing).

To my friends (you know who you are), thank you for being understanding, supportive, and so darn smart and funny.

Finally, I owe the heartiest portion of gratitude (and a long work-free vacation) to my one-in-seven-billion spouse, David Courage. Dave, you make me a better writer and a better person. Thank you for your very thoughtful (and very thorough) edits, insightful and clever comments, thousands of acts of support during this long process (while you were in graduate school and working, no less), and your overall caring presence. Thank you. I love you. Where should we go?

INDEX

TK

Index

Index

Index

Index

Index

Index

Index

ABOUT THE AUTHOR

TK